AI-Driven Healthcare: Innovations, Challenges, and Future Prospects

Harshit Kohli

Chitrangadaa Singh Kohli

Preface

There has been no greater impact on our world than artificial intelligence and healthcare together. We are witnessing how AI is altering how we comprehend, administer, and deal with medical care, from diagnostics to treatment personalization, but the speed of AI innovation also brings both unknown opportunities and issues. ***AI-Driven Healthcare: Innovations, Challenges, and Future Prospects*** came out of a significant need to chart this fast-evolving landscape with both a view of optimism about innovation and a critical eye toward ethical, technical, and practical challenges and inevitably shortcomings.

This book is the product of [has been produced on the basis of your motivation: e.g., years of research, clinical experience, or collaboration with technologists and policymakers]. Healthcare professionals, technologists, policymakers, and students can utilize it to conduct a survey of advancements such as predictive analytics, robotic surgery, and AI-enhanced drug discovery. However, the book also offers a direct examination of biases in algorithms, data privacy concerns, and the societal implications of entrusting care to machines.

We would also like to extend special thanks to our mentors, editors, and family for always supporting us in doing the right thing. Above all, thank you to the reader of these ideas. It is my intent with this book to give you the knowledge to be able to be a participant in co-creating an AI-augmented healthcare future in which advances in technology are complemented by human values that have stood the test of time.

Dedication

This book is dedicated to the myriad individuals who find themselves at the crucial juncture of humanity and technology within the realm of healthcare.

To the dedicated physicians, nurses, and clinical practitioners who welcome innovation while remaining steadfast in their commitment to compassion, who adapt to new systems without compromising their fundamental oath to heal, and who perceive artificial intelligence not as a substitute but as a supportive partner in their noble endeavors.

To the data scientists, engineers, and AI researchers who channel their intellect into addressing the most challenging issues in healthcare, who write code with a profound awareness of the lives their algorithms may impact, and who uphold a steadfast dedication to ethical progress. This dedication extends to the patients whose trust fuels advancement, whose data facilitates learning, and whose experiences serve as a poignant reminder of the significant purpose behind every technological leap in medicine. Additionally, to the healthcare administrators, policymakers, and regulators who navigate complex challenges to ensure that AI systems benefit all individuals equitably while protecting privacy and dignity. To the educators who are shaping the future workforce for a healthcare environment transformed by artificial intelligence, instilling both technical skills and humanistic principles. Lastly, to the trailblazers—both celebrated and those who remain in the shadows—who dared to dream of a future where human insight and artificial intelligence work in harmony to broaden the horizons of healing, alleviate suffering, and honor the intrinsic worth of every individual.

It is the hope that this book will aid in the thoughtful integration of AI into the intricate art and science of medicine, always prioritizing the service of humanity.

Acknowledgment

We extend our sincere gratitude to the many individuals who contributed to the creation of AI-Driven Healthcare: Innovations, Challenges, and Future Prospects. We are deeply thankful to the researchers, healthcare professionals, and technology experts whose insights and work continue to push the boundaries of artificial intelligence in medicine.

Special appreciation goes to our colleagues and collaborators who provided invaluable feedback and support throughout the writing process. We are also grateful to our family and friends for their patience and encouragement.

Finally, we acknowledge the rapid advancements in AI and healthcare that inspired this book, and we hope it serves as a meaningful contribution to the ongoing dialogue about the future of technology in improving patient care and medical outcomes.

Contents

Chapter 2: AI in Healthcare Operations and Management

Chapter 3: Challenges and Ethical Concerns in AI Healthcare

Chapter 2: AI in Healthcare Operations and Management — 57
Chapter 3: Challenges and Ethical Concerns in AI Healthcare — 104

AI and Technological Innovations in Healthcare

1.1 The Role of AI in Modern Medicine

Modern medicine today is in current transformational evolution in the application of artificial intelligence in terms of improving diagnostic accuracy, personalizing treatment, enabling faster drug discovery, monitoring patient parameters in a better way, as well as optimizing healthcare operations. Severe analysis of big medical data motivated using machine learning, deep learning, and natural language process (NLP) to find patterns and generate insight to support clinical decisions. AI models trained on medical images, patient records, and clinical data are used in diagnostics to tell that one is diagnosed with cancer, heart disease or he is afflicted with any neurological disorder. By using computer vision-based AI tools, tumors, fractures, and other types of abnormality on medical scans can be identified, outperforming humans in accuracy, and allowing for early diagnosis, and in turn, intervention (Bellec & Boyle, 2019)yet current models tend to focus on elementary cognitive processes. By contrast, video games have been designed to fully engage players, and require to constantly monitor the state of the game, in parallel to integrating strategic planning, decision making and taking action. While playing video games is hard, recent advances in artificial intelligence (AI.

There are more examples of AI improving personalized medicine, and analyzing genetic, clinical, and lifestyle data that are used to suggest personalized treatments. The machine learning models in pharmacogenomics help predict how specific drugs shall work in an individual and customize treatment plans as well as minimize adverse side effects of treatment. AI enables faster identification of potential drug candidates in drug discovery and simulates them against biological systems to lower the time and expense of bringing new treatments to the market. Moreover, AI is also used to repurpose drugs, that is, find new uses for existing drugs.

AI-powered wearable devices are used in patient monitoring by tracking a patient's vital signs like heart rate, blood pressure, and glucose levels, and monitoring early signs of health deterioration. They predict healthcare providers to potential emergencies improving response times as it relates to patient outcomes. Continuous analysis of patient data by AI-driven systems in intensive care units (ICUs) for detection of complications and minimal time required intervention. Remote monitoring platforms can replace access to doctors and save patients time for hospital visits.

The different functions of AI in healthcare operations include streamlining use of administrative tasks, optimizing resource allocation, and improving patient flow. Medical transcription, billing, and coding are automated by the NLP-based tools and it reduces the burden of the clerical staff in healthcare. Virtual Medical Assistants powered by AI give patients relevant medical information, help to schedule appointments, and answer patients' health-related questions; making the patients more engaged and satisfied with the services offered. AI models help in identifying patterns of disease, predicting outbreaks, and suggesting public health strategies in medical research and public health. In times of the COVID-19 pandemic, AI proved more than useful for modeling virus transmission, building diagnostic tests, and developing vaccines.

While very promising, both the privacy of data in use by AI in medicine, the potential for algorithmic bias, and the requirement of model transparency make this type of medicine still quite a challenge. For building trust among healthcare providers and patients, it is important to ensure that AI systems are compliant with data protection regulations, and allow the creation of explainable insights. For true AI to mature and become an integrated part of the fabric of the healthcare infrastructure, and the clinical workflows themselves, it requires an integration. And as AI advances, multi-modal AI, federated learning to bring more power to what AI can do for modern medicine.

1.1.1 AI vs. Traditional Healthcare Approaches

AI and traditional healthcare approaches differ significantly in how they handle diagnostics, treatment, and patient care. In diagnostics, AI models analyze medical images such as X-rays, MRIs, and CT scans with high accuracy, often detecting subtle anomalies that human experts might miss. In contrast, traditional diagnosis relies on the experience and expertise of doctors and radiologists, which can lead to variability and human error. AI also enhances treatment planning by analyzing genetic data, clinical history, and patient response patterns to create personalized treatment plans. Traditional treatment planning, however, follows standardized protocols and general clinical guidelines, which may not be as tailored to individual patient needs (Topol, 2019).

In drug discovery, AI accelerates the process by predicting molecular interactions, identifying potential compounds, and simulating drug effects, significantly reducing time and cost. Traditional drug discovery relies on a trial-and-error approach involving laboratory testing and clinical trials, which is time-consuming and expensive. AI also improves patient monitoring through wearable devices and predictive models that provide real-time health data and detect early signs of complications. Traditional patient monitoring, on the other hand, depends on periodic check-ups and manual assessments, which may delay the detection of health issues.

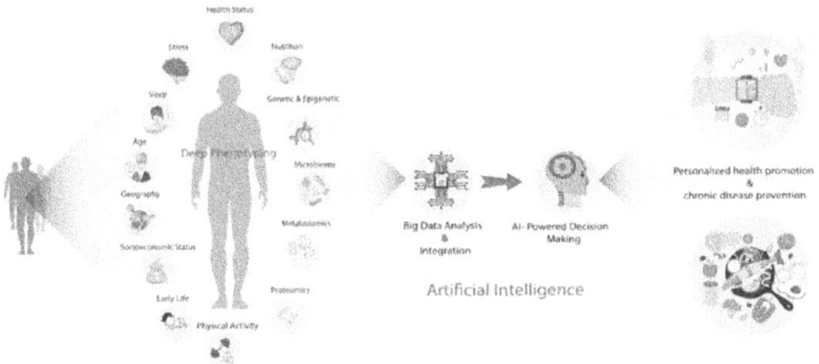

Figure 1.1: Predictive Analytics in Healthcare.

Source: (Subramanian et al., 2020)

Healthcare administration AI automates tasks like medical transcription, billing, and scheduling and makes it easier for healthcare staff as it reduces their burden. Administrative tasks for traditional systems need manual handling, and taking time, and prone to errors. Real time medical decision support systems using AI analyze the complex medical data and draft recommendations for treatment, while traditional decision-making involves the physician's experience and consulting the specialist. When it comes to patient engagement, virtual assistants and chatbots aided by AI support the patients regardless of the time by answering patients' questions and scheduling appointments. Traditional patient engagement is relying more on face-to-face consultations and phone-based communication and is limited by availability.

However, it is because of scalability that AI can run on large data sets and serve many patients simultaneously without human constraints. Traditional healthcare is limited by the shortage of healthcare professionals and sources. AI also helps public health and research by making predictions about disease outbreaks, optimizing resource allocation, and designing an optimal public health strategy based on epidemiological data. The methods for public health are historically oriented and efforts are made using historical data along with expert analysis, but these sorts of methods do not usually capture real-time patterns. Though, models using AI may not suffer these challenges, if they are trained on unbalanced datasets, they can suffer from or introduce bias. Likewise, traditional methods also vulnerable to human error and unconscious bias, which leads to decrease in accuracy of diagnosis as well as in treatment decisions. AI and traditional healthcare approaches have their own strengths and weaknesses however, one can expect tremendous benefit from merging these two aspects to better and effectively deliver healthcare.

1.1.2 The Evolution of AI in Medical Sciences

Over the past decades, the development of AI in medical sciences has been very advanced, and the way diseases are diagnosed, treated, and managed has changed. AI's foray in healthcare started with early rule-based expert systems in the 1970s and 1980s. MYCIN and other systems used predefined rules in diagnosing bacterial infections and recommending antibiotic treatments. However, these early AI systems were not significantly effective in real-world clinical settings since they were unable to process large datasets as well as adapt to new information.

Since the 1990s until 2000s medical AI had a turning point with the advent of machine learning (ML). Systems became able to learn from the data rather than only using predefined rules thanks to ML algorithms. By shifting this way, AI models were able to highlight the patterns in the complex medical data thus helping in better diagnostics and recommendations of treatment. For instance, AI had proved to be more effective than the traditional methods in analyzing medical images to spot cancer, heart disease, etc., with more accuracy. NLP also allowed AI to learn insights from unstructured clinical notes thereby making medical record analysis and decision support better.

In 2010s, introduction of deep learning accelerated the depth of AI's impact on medical sciences. Analyzing medical images, subject to genomics data or a patient history, the deep learning models (CNNs and RNNs) achieved great accuracy. And taking tumors, for instance, AI systems could now identify, detect, predict disease progression and genetic mutations with high accuracy. Instead of simply acting as another piece of data that could be effectively processed by humans, AI employed generative models to leverage its capacity to design novel drug compounds, to simulate molecular interactions, to optimize clinical trial outcomes in highly efficient ways; time and money spent on drug discovery was dramatically shielded.

AI's evolution went so far as to real-time patient monitoring and personalized medicine. Wearables with AI algorithms monitor vital signs and signal early signs of health decline to the healthcare provider in advance of complications. Today, AI has transformed the way chronic diseases, like diabetes or hypertension, are managed with the appropriate treatment plans given to each patient in accordance with patient-specific data. The emergence of federated learning enabled models not to compromise patient privacy by enabling them to be trained without having to be on centralized dataset.

Today, AI has reached further and further with the increase of multi-modal AI, combining text, images, and genomic data to produce sums that are coherent. Robotic surgery is also being aided by AI, as well as other administrative tasks that have been automated, along with supporting medical research with walked advanced data analytics. Albeit data privacy, algorithmic bias and model interpretability, AI is ready to be even more involved in medical sciences. AI in healthcare's future is to be integrated with artificial intelligence-enabled insights with human wisdom to make the healthcare system more accurate, efficient, as well as accessible.

1.1.3 AI in Enhancing Clinical Decision Support

It has improved healthcare professionals' decision support (CDS) with real-time insights advanced diagnostics and the best treatment plan. Part of the powerful AI algorithms, such as machine learning (ML), deep learning, and natural language processing (NLP), clinical decision support systems (CDSS) use to analyze vast amounts of patient data, medical literature, and clinical guidelines. CDS systems powered by AI help physicians in detecting patterns; predicting disease progression; and suggest evidence-based treatment options. These systems facilitate the reduction in diagnostic errors and effective care outcomes by supplying timely and accurate clinical recommendations (Rief, 2005).

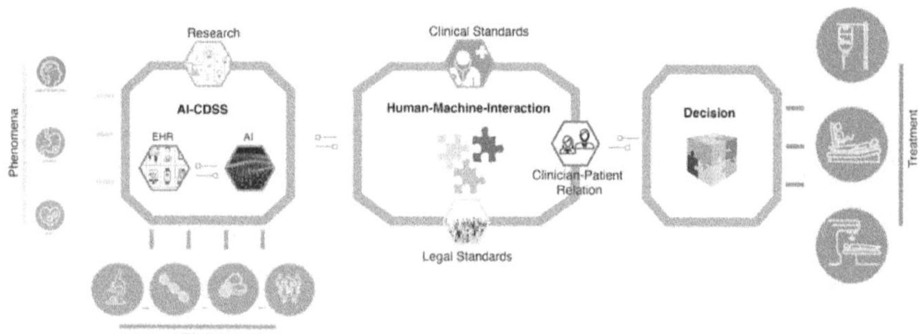

Figure 1.2: Clinical decision-making using AI-CDSS involves designing, generating, and analyzing data, focusing on human-machine interaction, and managing patient treatment outcomes.

Source: - (Bleher & Braun, 2022)

The added value of AI to CDS in the diagnostic will process medical images, lab results and patient records with an accuracy of high level.

With the help of convolutional neural networks (CNN) and similar models, diseases such as cancer, heart disease, and stroke can be detected from radiological and pathological images achieving better results than results of others. Human clinicians tend to miss some early signs of disease, while most AI models analyze complex imaging data to pick up some of the things that happen early and identify them. For example, AI applied on mammography to find earlier-stage breast cancer has also been used to improve survival rates.

Prediction of disease progression depends on a patient risk assessment, generated from AI-driven predictive models to help clinical decision making. Based on historical patient data, machine learning models can predict the individuals at high risk for sepsis, heart attacks, and renal failure, and early intervention can be done which can help in reducing the complications. Personalized medicine is also supported by AI that suggests personalized treatment plans aimed at a patient's gene make up, clinical history, and reaction to past treatments. For instance, AI system based on pharmacogenomics considers genetic data to predict how patients will react to a particular medication so that it can be more precisely determined and medication dosage established.

CDS is improved by NLP in that it can take insights from unstructured clinical notes, medical literature, and research papers. Knowledge that AI systems can locate relevant studies, clinical guidelines or drug interactions allows healthcare providers up-to-date information to assist in decision-making. CDS tools based on AI are integrated into electronic health records (EHR) to alert and suggest real-time during patient care. For instance, AI can identify potential drug interactions, suggest alternative treatment options and nonemerging medical diagnoses that can result in fewer human errors and enhanced patient safety.

Additionally, AI helps in automating workflow by streamlining such tasks, such as chart review, documentation, and order entry by removing extra administrative work. An AI assists the clinicians to do what they are trained to do, and frees them from the administrative burden. In addition, the accuracy and relevance of an AI-based CDS system increase with continuously learned and perfected AI from new data. Despite these challenges, keeping their data private, the AI in

CDS encounters problems such as data privacy and algorithmic bias, issues of algorithmic bias, and model transparency that are equally important when it comes to trust and effectiveness. With advancing AI technology, the role of AI integration into clinical decision support systems will also significantly contribute to the improvement of diagnostic precision, treatment outcomes as well as in efficiency of healthcare.

1.2 Advancements in Diagnostics and Medical Imaging

AI has also advanced disease detection, diagnostics, and medical imaging greatly improving how fast and how accurately it can pinpoint diseases and how quickly and accurately planners can plan treatments. X-rays, CT scans, and MRIs were traditionally largely dependent on the human eye and interpretations could vary along with being badly missed. However, AI is conducting this process using deep learning and computer vision. Convolutional neural networks (CNNs) have been proven to have a great potential in analyzing a lot of complex medical images and finding patterns, types of abnormalities that usually get overlooked by human radiologists. The use of AI-based equipment for scanning or reading mammograms or any other related cancer has been proven to be very accurate in detecting that particular cancer such as breast cancer, lung cancer, and skin cancer. For example, Google's DeepMind built an AI model that was able to detect breast cancer as good or even better than human radiologists in diagnosing breast cancer from mammograms. For instance, it has helped improve:

1. Cardiology diagnostics through the identification of arrhythmias in electrocardiograms (ECGs).

2. Prediction of heart failure from review of patient history. AI models have been very successful at diagnosing Alzheimer's or Parkinson's disease in neurology by detecting little changes in brain structure in an MRI scan for early intervention and better patient outcomes(Yoon, 2021)or professional handbook, on topics at the interface between machine learning, spatial statistics, computer simulation, meta-modeling (i.e., emulation.

Furthermore, there has been tremendous advancement on the front of AI-based pathology and genetics diagnostics. Whole Slide Imaging uses deep learning on images of tissue samples to automate the assessment of the tissue sample with high precision and to detect cancerous cells, or tissue abnormalities, for example It allows AI to take data from more than 1 source, which is imaging, lab results, patient history. These are processed by the AI to make the diagnostic more accurate and hence help in further decision making. This is to give patient a bit better, more comprehensive understanding concerning the state of his health, and to perform more accurate complex diagnoses. Progress of the diagnostic process is further streamlined from real time analysis to further automated reporting and their systems reduce the burden on health professionals and enhance the care of patients. In return, AI has also worked to utilize genetic information to study these mutations of a certain disease that, in turn, are used to get more specific treatments. However, data privacy, algorithmic bias, and the irrepressible appeal of the necessity for regulatory oversight remain. AI will have a longer application in diagnosis and medical imaging as AI's role in detection of early disease, patient care, and overall efficiency of healthcare.

1.2.1 AI in Radiology: X-rays, CT scans, and MRIs

Radiology is a central part of human body structure and function imaging and modern medicine is based on this fact. X rays, computed tomography (CT) scans and magnetic resonance imaging (MRI) are the three most commonly used imaging modalities used for the diagnosis or follow-up of a medical condition and each have their own advantages. These challenges have been higher demand for imaging services, complicated cases, inadequate radiologists and have been in concurrence with the need to deliver timely and accurate diagnosis.

Artificial Intelligence (AI) presents machine learning (ML) and deep learning (DL) as it brings great changes to radiology as this field is fully automated, helping analyzing images, while improving the diagnostic accuracy and the radiological workflow. AI models can be used for Medical Image Analysis to detect subtle abnormality and give clinical support in making decisions in the presence of complex pattern on medical images. They enable earlier diagnosis and earlier development of a treatment decision and better patient outcomes.

Using three major imaging modalities (X-rays, CT scans, and MRIs) — this section shows how, in each case, AI has been conducting benefits, challenges, and clinical applications in each of the cases of AI driven solutions.

AI in X-ray Imaging

A cornerstone of medical diagnostics is x-ray imaging, after Wilhelm Roentgen's discovery in 1895 that can be done easily and cheaply, to see bones, soft tissues and internal organs. For such a simple and popular device, X-rays require a high degree of understanding from its interpreters, because such abnormalities as a small fracture, early-stage cancer and early changes in lung disease are easily missed. X-ray analysis has become a new important tool, which combines artificial intelligence (AI), machine learning (ML), and deep learning (DL) models, especially convolutional neural network (CNN), with the aim to increase the diagnostic accuracy and efficiency. These methods are trained on large datasets consisting of annotated X-rays and have shown the capability to identify abnormalities with the degree of precision comparable to, and sometimes even better in some cases, than human radiologists.

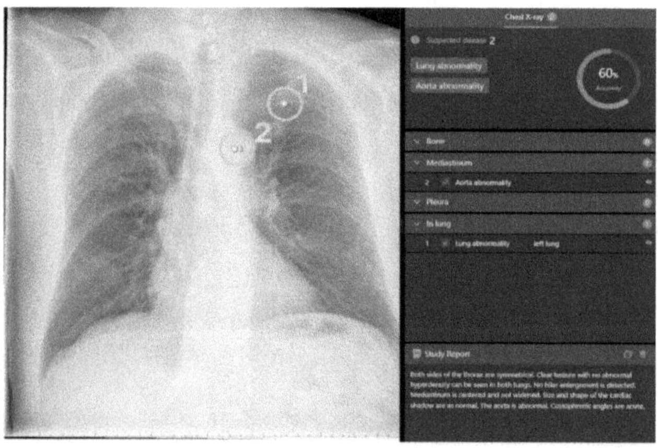

Figure 1.3: Chest radiograph analyzed with AI. AI-generated report based on a routine chest radiograph marking two findings together with their precise locations: one lung abnormality (1) and one aortic abnormality (2). All findings are listed on the right side, along with an overall abnormality probability score of 60%. Below the list is a concise report created by the software, similar to a radiology report.

Source: - (Kromrey et al., 2024)

Lung diseases such as pneumonia, tuberculosis, and lung cancer have been shown to have been very well detected by AI for lung fields texture, shape, and density. Specifically, AI models have surpassed human radiologists in identifying pneumonia, especially in mild or atypical cases, as well as in their ability to detect early signs of tuberculosis and distinguish COVID-related pneumonia from other kinds. AI models have been able to determine fracture and joint abnormalities with high accuracy in musculoskeletal imaging. Anatomical structures present on the wrist and ankle, as well as on the ribs and even on young pediatrics, for instance, growth plates, are complex to identify, therefore, AI algorithms have contributed to better detection rates and classification of the severity of injuries. AI has brought to cancer screening great advancement in early detection of lung and breast cancers by detecting very small abnormalities like pulmonary nodules and microcalcifications that typically human radiologists cannot detect. Detection is not only enhanced, but we also give probabilistic scoring and risk assessment to help decide on further evaluation.

Beyond detection, AI allows for automation of AI, triage, and measurements, and controls quality. Triage systems based on AI can sort out critical cases like pneumothorax or lung collapse to make them priority cases that can be attended to immediately. Bone density, joint space narrowing, and lung volume can be measured by automated measurement tools for the diagnosis of osteoporosis, arthritis, or chronic obstructive pulmonary disease (COPD). Also, AI helps to improve image quality by detecting poor positioning or inadequate exposure and moving on to a repeat scan if needed. Radiologists receive real-time suggestions from AI-based decision support systems, highlight the regions of interest, generate structured reports, and hence Cognitive load reduces and diagnosis becomes consistent. AI automates complex tasks and increases diagnostic accuracy thereby reducing the number of simple cases, which can be handled by radiologists, then focusing on complex cases, which would subsequently result in a better chance of patient survival and accelerated delivery of healthcare to patients.

AI in CT Scan Analysis

CT scans generate very detailed cross-sectional images of internal structures and are used to image generally complex anatomical regions like brain, chest, abdomen, and the cardiovascular system. Large volumes of high-resolution data are produced by CT imaging, which is very time-intensive and requires significant expertise. As an innovative approach to automation and improvement of CT analysis, AI has been playing a significant role in generating diagnostic accuracy as well implications for reducing interpretation time. The existence of deep learning networks such as convolutional neural networks (CNNs) or transformer-based architectures, and the availability of large amounts of data, are the two building blocks for AI-based CT analysis. However, AI can sense subtle abnormalities and automate complex measurements and predictions. This helps greatly in clinical decision-making.

What CT imaging shows is that AI is remarkably successful at detecting stroke and other neurological conditions. AI models can help identify blocked blood vessels and reduced blood flow in ischemic stroke cases, within minutes, thus aiding in faster decisions and better patient endpoints. Because AI can detect subtle changes in density and texture in brain tissues, even in cases with mixed densities that can obscure the bleed, it can detect brain hemorrhages for hemorrhagic stroke. Stroke detection tools based on AI are now integrated in emergency departments' workflows and these are leading to faster triage and more accurate diagnosis, and consequently shorter time to treatment and hence higher recovery rates.

Also, AI is extensively utilized in oncological imaging for tumor detection and characterization. AI models are able to identify small nodules in lung cancer screening with high sensitivity and specificity and to classify them according to malignancy risk. AI can be used to detect lesions in complex abdominal structures and assess the lesion's likelihood of being malignant, based on texture, vascular patterns, and shape, for liver and pancreatic cancer. AI-based virtual colonoscopy systems have become useful in colorectal cancer to detect polyps and early-stage tumor that reduce the invasive procedure along with early diagnosis and intervention.

AI has also been used to analyze CT for cardiovascular imaging. Using AI models, coronary artery calcium scores can be measured to indicate plaque burden and the starting point of atherosclerosis, these signs of disease can be detected, and other indicators of heart health and disease risk can be quantified. AI-based automated cardiac function analysis of LVEF, chamber volume, and thickness of myocardial wall helps in the disease diagnosis and treatment of heart failure and valvular heart disease. The 3D reconstructions of coronary arteries provided due to AI are used in the surgical planning and risk assessment process for the betterment of patient outcomes.

In many trauma and emergency ambulances, there has been the adoption of AI-based CT analysis systems to rapidly detect life-threatening injuries. Being able to identify internal bleeding with a bit of hyperdense area which reflects the accumulation of blood in the brain, chest, and abdominal area can allow the response to be done quickly with surgical or its corresponding radiology intervention. In addition, AI is able to detect organ lacerations in the liver, spleen, and kidneys, and provide more information about the size, depth, and severity of an injury. In addition, the AI models are capable of detecting such fractures with spinal cord injury, as well as generating 3D reconstructions useful in surgical planning. This real and fast analysis helps in making a quick decision and helps to save the life of patients in critical care situations.

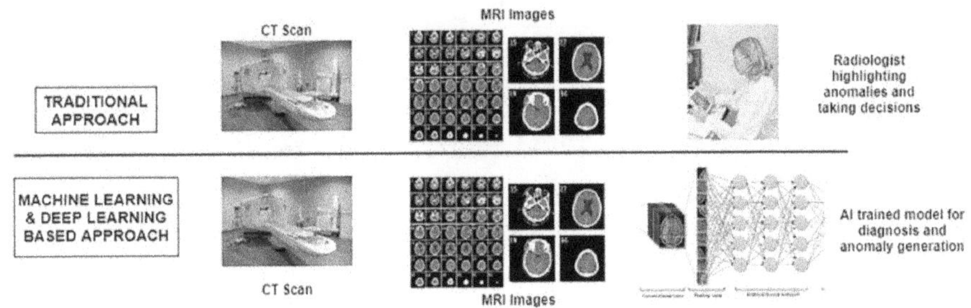

Figure 1.4: Radiologist vs. AI for CT and MRI Interpretation

Source: - (Ahsan et al., 2023)

CT analysis constitutes increases in diagnostic accuracy, decreases in human error, and improvements of clinical workflows. AI enables radiologists to process larger volumes of data more efficiently and better for the patients which results in providing better patient care and treatment outcomes. With the development of AI models, future CT imaging and clinical decision-making will further be revolutionized, as the number of data we can analyze are becoming more complex and in multi modals.

AI in MRIs

Magnetic Resonance Imaging (MRI) is a very sophisticated picture-making technique that creates highly detailed pictures of internal structures with strong magnetic fields and high-frequency radio signal pulses. Unlike X-rays and CT scans, MRI does not employ ionizing radiation and is, therefore, safer for repeated imaging, especially in those who are born or about to conceive, such as children. MRI aligns the hydrogen protons in the body's tissues by using a strong magnetic field. Protons are exposed to radiofrequency pulses that generate signals that are captured and computer algorithmically processed into detailed images. This results in outstanding visualization of soft tissues such as brain, spinal cord, muscles, ligaments, and internal organs, and MRI is the gold standard for diagnosis of neurological, musculoskeletal, and cardiovascular conditions(Jin, 2018).

Motor Design Sensors: Using professional-grade sensors, we were able to convert the FlexTouch Sensors into beacons for private charging. Manual review of hundreds of image slices is needed for traditional MRI and it is time-consuming and subject to human error. With great success, deep learning models, i.e., convolutional neural networks (CNNs) and recurrent neural networks (RNNs) have been able to automate MRI interpretation, and hence improve diagnostic accuracy. The subtle abnormalities that cannot be seen by human eyes are detectable by AI algorithms, and the anatomical structures can be quantified and radiologists can be helped to make real-time decisions. AI automates routine tasks resulting in reduction of the cognitive load on radiologists and boosts the speed of reaching clinical decision.

AI has improved significantly on the diagnosis of neurological disorders including stroke, multiple sclerosis (MS) and brain tumors in brain imaging. AI models trained on large MRI datasets can predictions of severe ischemic stroke and distinguish them from voids in infarctions. AI can accurately localize hemorrhagic bleeds and estimate its volume, useful for surgical planning for hemorrhagic stroke. Recently, AI models have been shown to be sensitive to detecting white matter lesions as well as to their progression in MS. Additionally, models have been trained to spot brain tumor patients based on their image shape, texture, and vascularity, and with the help of AI. Precise measurement of tumor size and growth can be obtained by automated tumor segmentation and this can be used within treatment planning and monitoring.

AI models have also been used for improving the detection and subclassification of joint abnormalities, and ligament tears, and spinal cord injuries in musculoskeletal imaging. In fact, AI is able to analyze knee MRI scans and identify which meniscal tears and cartilage damage are present, and helps with automated grading to assist orthopedic surgeons. AI applied to Spinal MRI has boosted the exploration of disc herniation, spinal stenosis and vertebral fracture and labelling and automated 3D reconstructions help with surgical planning. With AI enabled MRI analysis the injuries in muscles and tendons can be detected early on and targeted rehabilitation and decrease the risk of re-injury can be enforced at an early stage in sports medicine.

Still, cardiac MRI (CMR) is another area where AI has seen great development. The cardiac function parameters like the left ventricular ejection fraction (LVEF), the myocardial wall thickness and the stroke volume can automatically be quantified by AI models. Current AI based tissue characterization models have demonstrated very high accuracy on detecting myocardial infarction, fibrosis, and inflammation. More accurate assessment of the structural and functional abnormalities can be achieved through automated segmentation of heart chambers and coronary arteries. Early diagnosis of cardiomyopathies, valvular heart disease, and arrhythmogenic disorders is possible in instances when AI-enhanced CMR analysis is utilized to assist cardiologists in providing their curative treatments, minimizing the initial cardiologists' load.

MRI acquisition and reconstruction have also been sped up by AI. MRI scans typically take anywhere from 30 minutes to over an hour, during which the patient contemplates his or her own demise as a result of exposure to metal, consumes unknown amounts of healthy but hyper anagrammed fluid, and/or suffers motion artifacts that make positing the hand obvious difficult for the technician. Utilizing this compressed sensing to reduce the required scan times by up to 50% has been used to develop AI based acceleration techniques, e.g., parallel imaging. Therefore, deep learning-based reconstruction algorithms can generate high resolution images from under sampled data leading to further reduction of scan time and better patient comfort. Motion artifacts can also be automatically corrected for by AI models, which can then also optimize image contrast to optimally increase the clarity and allow for better diagnosis than before of MRI scans.

Whole body MRI has proven to be successful in oncology for staging and monitoring of cancer using AI. AI models can be identified as highly sensitive and specific for finding metastatic lesions in the liver, lungs, and bones. At present, there is AI based multiparametric MRI analysis for early detection, localization of tumors and risk stratification for prostate cancer. In terms of breast MRI analysis, promising AI application include the ability to detect invasive and noninvasive lesions with less false positives. Automated tumor volume tracking is called because oncologists can monitor the response to treatment and adjust therapy.

AI based workflow automation in MRI has eased the manual work in radiology practice and improved consistency. This means that the AI powered triage systems can automatically prioritize the MRI cases based on the seriousness of abnormalities detected, such that critical cases can be given immediate attention. AI generated structured reports standardize diagnostic interpretation, reduce interobserver variability, and improve communication among all the teams of the healthcare. Quality control algorithms will detect image artifacts, patient motion or incorrect positioning and would prompt them to repeat scans when necessary.

AI has been incorporated into MRI analysis and increases diagnostic accuracy, speeds up the viewing time of MRI and improves patient outcomes in several medical fields. Having the ability of AI to analyses complex patterns and huge dataset has empowered the radiologists with ability to make fast and more confident decisions. As AI models incorporate multimodal data (such as MRI, in conjunction with clinical data, underlying genetics), they will continue to enhance the capacity to predict and the associated drivers of a magnitude of individualization in personalized medicine, and targeted therapies. AI changes MRI far more fundamentally than analysis of it, and even diagnostic imaging and patient care are being redefined by AI.

1.2.2 AI for Early Disease Detection

Early detection of disease is important to optimize patient outcome, lower health care costs and enhance the results of appropriate treatment. As in many other medical areas, traditional diagnostics still often relies on detecting symptoms and standard imaging and/ or laboratory tests that often only detects the disease when it has progressed to a considerable degree. For finding out the diseases at early stages, Artificial Intelligence (AI) have evolved as a powerful medium to address the complexity of these medical data (imaging, biomarkers, genetic information, patient history) to aid in disease diagnosis. Due to the subtle signs that AI models like deep learning and machine learning algorithms can uncover that are invisible to human clinician eye, AI models can help detect earlier and cause effective action. This provides better and quicker analysis on a variety of decisions in the area of healthcare helping its providers to make more informed decisions that help reduce diagnostic errors and improve the care they give the patient(Wager et al., 2009).

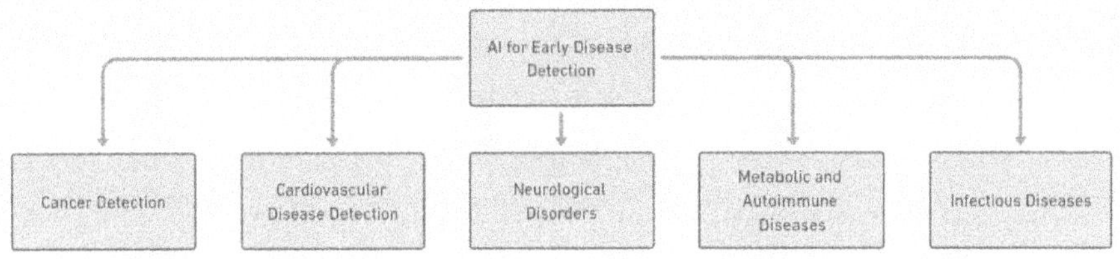

Figure 1.5: Early Disease Detection with Artificial Intelligence

Source: - (Self-generated)

a. **Cancer Detection**

AI has great promise in oncology to detect cancer earlier in a disease, increasing the odds of curing it and prolonging survival. Mammograms, CT scans and MRI scans are large datasets of which AI powered models can be trained to identify small tumors and microcalcifications showing that they can be sensitive and specific in identifying breast cancers. AI models have also been successful in finding pulmonary nodules on low dose CT scans and in predicting their malignancy risk based on growth pattern, size, shape and morphological features in lung cancer screening. Just as, AI algorithms have been used for identifying early indications of colorectal cancer through searching polyps and lesions in virtual colonoscopy scans and increasing the rates of early diagnosis and necessitating fewer invasive procedures. Similarly, AI based cancer detection systems also assists in detecting metastasis and help the oncologists in determining the extent of disease spread and consequently designing a proper treatment plan.

b. **Cardiovascular Disease Detection**

AI has also achieved great progress in early detection of cardiovascular diseases, which are among the top causes of mortality worldwide. Electrocardiograms (ECGs), cardiac MRI, and CT angiography can provide an insight into the early signs of coronary artery disease, arrhythmias, and heart failure that machine learning models can look into. Coronary artery calcium scores and plaque burden can be analyzed based on an AI, which can lead to predicting the risk of Myocardial infarction and giving decisions about preventive treatment strategies. AI models can detect atrial fibrillation and other cardiac abnormalities that carry stroke risk to be used in identifying patients that may go on to have a stroke before they actually occur. In addition, blood flow patterns, heart valve function, and myocardial strain-based condition can also be assessed by AI. With the ability to process large quantities of patient data in real time and in real time, AI can identify what is not possible in a couple of minutes.

c. **Neurological Disorders**

Neurological disorders are characterized by early detection difficulties because Alzheimer's and Parkinson's are both subtle, progressive diseases. As mandated by health services, many require specialized healthcare in specific hospitals. And unfortunately, most of these hospitals lack accessible technology that could identify and manage these complex health issues by detecting early structural or functional changes in brain tissue linked to neurodegenerative diseases, such as Alzheimer's and Parkinson's, and other health problems, for example, type II diabetes. Early diagnosis of brain atrophy, reduced metabolic activity, and protein deposition can be accomplished by AI and relevant therapies can be targeted before severe cognitive decline takes place. AI also has been trained with the goal of diagnosing Alzheimer's disease through scans, e.g. being able to identify amyloid plaques and tau tangles in subjects' brains, hallmarks of Alzheimer's disease. AI based gait analysis and speech pattern recognition systems assist in early detection of motor and non-motor symptoms in Parkinson's disease, and higher chances of earlier therapeutic interventions. In addition, AI can look at genetic and biomarker data to flag people at risk of neurological disorder development who can be the subject of individual prevention strategies.

d. **Metabolic and Autoimmune Diseases**

Detection of metabolic and autoimmune diseases (both have nonspecific symptoms and complex disease mechanism) has also been promising for AI. In addition to retinal images, the AI models can detect early signs of diabetic retinopathy (DR) in diabetes patients before they are aware of the disease using retinal images. AI analysis of Electronic Health Record can predict the onset of type 2 diabetes by looking for the pattern of blood glucose and insulin resistance as lifestyle factors. Similarly, AI can help in diagnosing autoimmune diseases such as rheumatoid arthritis and lupus based on joint imaging, inflammatory markers as well as genetic data to forecast future appearance of early signs of tissue damage and dysfunction of the immune system. AI models can also foresee if an autoimmune condition is about to flare up and use that to fight the disease and reduce its progress. AI can be used towards metabolic syndromes such as obesity, hyperlipidemia jointly with lifestyle, genetic and clinical data to provide personalized health recommendations and early lifestyle interventions.

e. **Infectious Diseases**

AI models have been used to find early outbreaks of disease, to track progress in disease, and to predict a rise or fall of infection rates in infectious disease. It has turned out that AI algorithms examining chest X-rays and CT scans are very accurate in spotting the early signs of pneumonia and respiratory infections, which include COVID19. Most of the models of natural language processing (NLP) can monitor global health data, news reports and social media to detect clusters of disease activity and anticipate outbreak and prompt the rapid response to public health and resource allocation. Use of AI for diagnostic tool can distinguish bacterial from viral infections, helping in the better stewardship of antibiotics and reducing the risk of antimicrobial resistance. Nowadays, AI models were crucial in keeping track of the patient deterioration, predicting oxygen need, and triaging critical patients during the COVID 19 pandemic. Real time monitoring of infectious disease symptom; the integration of AI with wearable health devices has helped to integrate real time monitoring of infectious disease symptom, and thus provide faster intervention and better patients outcome.

1.3 AI in Drug Discovery and Development

AI in drug discovery and development is the use of artificial intelligence and machine learning to expedite drug discovery and development by planning, designing, testing and approval of the new drugs. Traditional drug discovery is tedious, expensive, and time consuming (in the order of 10–15 years, sometimes billions of dollars expended) for a single drug from start to finish. As it has completely transformed this landscape, AI has been able to achieve the same feat with drug discovery and development, achieving significant improvements in efficiency, accuracy and success rates(Gupta et al., 2021).

The size of these biological datasets and the number of potential drug candidate programs makes manual analysis infeasible. The machine learning algorithms are also used to find out the biomarkers, predict a development of disease, and personalize the treatment for a personalized medicine.

Once it is in the drug development phase, AI driven simulations can predict the pharmacokinetics and pharmacodynamics (those activities of absorbing, distributing, metabolizing, and eliminating of drugs) without the expense and time of laboratory experiments. AI is also used in designing and optimizing clinical trial by identifying what cohorts are most suitable patients and what their outcomes will be by genetic and environmental factors.

1.3.1 Accelerating Pharmaceutical Research with AI

This has been a long, complex and expensive process that takes a decade or more and billions of dollars' worth of costs to get a single drug to market. Challenge in the pharmaceutical industry is high failure rates and growing complexity of the disease. Today, artificial intelligence (AI) is helping to shape the scene of the pharmaceutical research – by ensuring greater efficiency, accuracy and success in the process of drug discovery and development. Researches are using AI driven solutions to analyze enormous biological and chemical data, to identify promising candidates and to streamline drug development pipeline. It is particularly important for reducing costs and developing timelines, as a fuel engineer would say, and importantly, for meeting critical urgent medical needs and improving patient outcomes.

At the early stage of drug discovery, AI is improving the target identification and screening. The usual method of identifying disease targets is through years of complex biological research. AI enabled models, especially based on machine learning (ML) and deep learning, can analyses the genomic, proteomic, and clinical data at a large scale and can discover the targets of the disease with higher accuracy and speed. Using AI to recognize patterns in biological data allows researchers to predict the most probable proteins or genes to participate in disease mechanisms, and how they might react to the intervention medication.

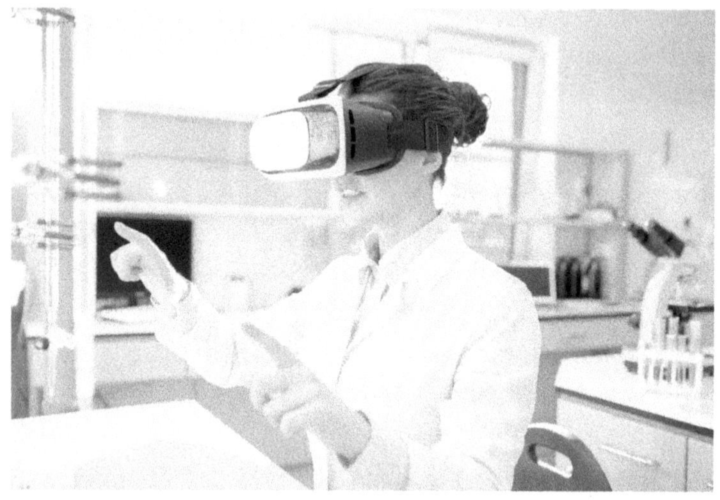

Figure 1.6: Virtual Reality in the Lab.

Source: (Hno & Gy, 2016)

AI also shortens the screen and optimize compound screening and optimization process. Currently, high through put screening (HTS) techniques have suffered from constraints on time and cost. Using AI, virtually millions of compounds can be screened, their binding affinity can be predicted, and predicted modifications to increase efficacy and decrease toxicity suggested. In light of these generative models (generative adversarial networks (GANs) and reinforcement learning algorithm) are being used to design new molecular structure that has desired pharmacological properties. This approach does away with the need to actually physically synthesize and test through many compounds, thus saving on laboratory costs and development time.

The application of AI in predicting interactions of drugs with targets in pharmaceutical research is one of the most impactful uses. In fact, through convolutional neural convolutional neural networks and graph neural networks, for example, AI models were able to analyze, and predict the interaction of the drug's molecular structure, with the biological target. Moreover, these models can help predict off target effects so that adverse side effects do not occur during development. Furthermore, it can also predict bioavailability and metabolic pathways of drug candidates to modify the molecular structure so as to enhance absorption and efficacy.

AI is also driving innovation in drug repurposing—identifying new therapeutic uses for existing drugs. This approach significantly reduces the time and cost associated with developing new drugs since the safety profile of existing drugs is already well-established. AI models analyze existing clinical data, drug interaction databases, and molecular structures to identify potential new applications. For instance, AI was instrumental in identifying potential treatments for COVID-19 by analyzing existing antiviral drugs and their mechanisms of action.

Preventing drugs causing undesirable effects and toxicity will become less expensive and less risky with the use of more efficient, faster, and cheaper AI that predicts potential adverse effects and toxicity profiles early during development. Chemical structure and biological response datasets, available in large size, are a potential source of information provided from deep learning models trained on them, in order to identify structural features associated with toxicity. Researchers can then modify or remove problem compounds or alleviate the need for improvements in later increasingly costly development stages.

AI is also benefitting pharmacovigilance, the process of monitoring and assessing the drugs' safety after approval. Using the rules, NLP algorithm can analyze medical reports, electronic health records, social media data among others to detect emerging safety issue and adverse drug reactions in real time. Such an approach is proactive as well, enabling the pharmaceutical companies to respond quickly to potential safety issues and protect the patients at the same time without facing the costs of recalls or enormous legal charges.

It is also helping to determine patient specific drug responses and drug metabolism variation due to genetic and environmental factors. It helps researchers to develop more specific therapies and consequently improve broad treatment outcome and reduce side effects. AI can help pharmaceutical companies to improve the precision and improve the effectiveness of new drugs, while at the same time bringing down the time and the costs to market.

Finally, AI facilitates drug discovery, increases the rate of drug–target interactions and drug safety. AI's capacity to analyze and process large data sets, to discover patterns hidden in the text, and to simulate the intricate biological interactions can change drug development, and then release to the market completely. Because this also helps pharmaceutical companies save money, reduce development time, and reduce the cost of pharmaceuticals for patients, and increase patient outcomes by providing safer, more effective drug treatment, it has wide support.

1.3.2 AI-Powered Clinical Trials and Patient Selection

The clinical trials are one of the most time consuming and expensive phase of pharmaceutical research, only approx. 10% of drug candidates make it to the market. AI is changing clinical trials by speeding up recruiting patients, better trial design and predicting outcomes. One of the main problems in selecting a patient for a clinical trial is that usually treatment response diversity is a costly discovery within each separate trial. The ability in large scale healthcare databases, genetic information, and electronic health records (EHRs), to do the AI driven models based on machine learning (ML) and natural language processing (NLP) to identify patient cohort that are most likely to show positive responses to the drug candidate. It significantly increases the trial success rate compared to classical approaches, and lowers the patient recruitment time and cost. Adaptive trial designs are also enabled by AI through which protocols are adjusted in real-time with respect to interim results. For example, AI can tell if a trial is heading towards efficacy or safety issues early, so researchers may change the dosing levels or rig the trial without restarting the trial(Fisher, 2009).

Likewise, AI is helping in monitoring, compliance, and data analysis aspects of clinical trials. The real time patient data which can be collected by such wearables and remote monitoring tools involving AI is the vital signs, activity levels and medication adherence. This data is analyzed by the AI models to identify patterns that may suggest adverse effects or treatment success, and alert can be taken in a timely manner. It also facilitates predicting outcomes of impending trial trials by analyzing historical trial data and patient response data. This enables sponsors to take a more accurate decision about whether or not to continue, modify or terminate a trial early therefore saving the resources and minimizing financial risk. In addition, data collection and analysis as a result of AI driven automation results in fewer human errors and accelerated reporting, further leading to improved overall accuracy and efficiency of a trial. AI is enhancing patient selection precision, improving trial design and increasing real-time monitoring for clinical trials while increasing success and decreasing costs for the trials, thus speeding up the process of new treatments going on the market.

Benefits of Using AI in Clinical Trials

Using AI for clinical trials offers several advantages that help enhance the accuracy, efficiency, safety, speed, and overall success of the drug development process. Mentioned below are some of the many benefits of AI in clinical trials.

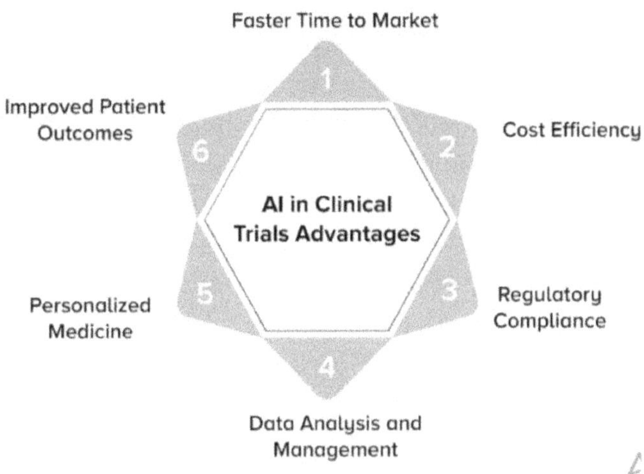

Figure 1.7: AI in clinical trial advantages

Source: (Askin et al., 2023)

- **Faster Time to Market:** One of the most obvious benefits of AI In clinical trials is the automation of labor-intensive and time-consuming tasks with remarkable speed and precision. For example, AI can swiftly analyze vast datasets, match patients to clinical trial criteria, and process complex scientific data, tasks that human researchers would take much longer to perform. As a result, R&D teams can expedite the drug development process, bringing potential treatments to patients more quickly.

- **Cost Efficiency:** By automating various research and development processes, AI can reduce the need for extensive manual labor and repetitive tasks. This leads to cost savings in terms of labor, resources, and operational expenses. Furthermore, AI can identify and prevent inefficiencies in clinical trials, reducing the risk of costly protocol amendments and ensuring that resources are allocated more efficiently.

- **Regulatory Compliance:** AI for clinical trials can also assist in maintaining compliance with regulatory standards by providing real-time monitoring, documentation, and audit trails for clinical trial data and processes. It ensures that the R&D team remains aligned with regulatory requirements, minimizing the risk of costly delays or healthcare compliance issues.

- **Data Analysis and Management:** The large amount of data generated in clinical trials can be overwhelming. AI can quickly analyze and organize the sheer volume of data and identify patterns that human researchers would take much longer time to spot or might overlook sometimes. Artificial intelligence in clinical data management helps the R&D team quickly access organized data, which saves time on manual data management and reduces the risk of data errors.

- **Personalized Medicine:** The use of AI and machine learning in clinical trials has also revolutionized personal medicine. Every patient deal with unique needs and complexities, making it challenging to test the treatment efficacy. AI can play a pivotal role in pinpointing particular patient groups that are most likely to benefit from a specific medication based on factors like genetic profiles and lifestyle, making personalized medicine a reality.

- **Improved Patient Outcomes:** Applications of AI in clinical trials also help tailor treatments to patients by identifying biomarkers, predicting treatment responses, and optimizing trial protocols. This patient-centric approach enhances the likelihood of successful outcomes for trial participants. Patients receive treatments that are more likely to be effective for their specific conditions, leading to better clinical responses and quality of life.

1.3.3 AI in Drug Repurposing for Existing Diseases

Drug repurposing using AI has become popular because it is helping identify new therapeutic application for existing drugs, thereby cutting down the time and cost needed to develop new therapies. Repurposing of off the shelf drugs represent a faster and cheaper solution to traditional drug discovery because safety profiles and pharmacokinetics of off the shelf drugs are already known. Biomedical data, including clinical trial results, molecular structure, protein interaction, disease pathway is vast and could be analyzed by AI models, especially deep learning and graph neural network for discovering unexpected relationships between drugs and diseases. This being recognized gives the AI the ability to predict how drugs already exist might go well with other biological targets, Finding new indications for treatment. One of those is the use of AI in analyzing antiviral mechanisms of drug action that has led to the rapid identification of potential drugs for COVID 19, without the use of traditional drug development timeframe.

By adding AI to drug repurposing, precision medicine is also gaining productivity through discovering patient subpopulations that could derive the most from current treatments. Machine learning models can be used to predict how a patient group will respond to a drug if it is treated with some drug. Such a targeted approach will form more individualized treatments as well as better patient outcomes. In addition, the application of AI makes it possible to quickly scan hundreds of thousands of drug compounds and to virtually simulate the drug target interactions, which can assist with developing and ranking the repurposing candidates. With the help of NLP algorithms, scientific literature, patents and medical reports can be used to extract valuable information, which in turn speeds up the discovery of new drug applications. AI enabled drug repurposing allows companies to reach the market faster with therapies at much lower cost and in much shorter time, which means delivering more patients efficient therapies for unmet medical needs.

1.4 Robotics and Automation in Surgery

Robotics and automation are revolutionizing the field of surgery by enhancing precision, improving patient outcomes, and reducing recovery times. Surgical robots, such as the da Vinci Surgical System, have become widely used in minimally invasive procedures, allowing surgeons to perform complex operations with greater accuracy and control. These systems use robotic arms equipped with advanced sensors and cameras to provide high-resolution, 3D views of the surgical site and allow for precise movements beyond human capability. AI-driven automation further enhances these systems by providing real-time feedback and suggesting optimal surgical techniques based on historical data and patient-specific anatomical details. This increased precision reduces the risk of complications, minimizes tissue damage, and shortens recovery times, leading to improved patient satisfaction and better overall surgical outcomes (Maibaum et al., 2022).

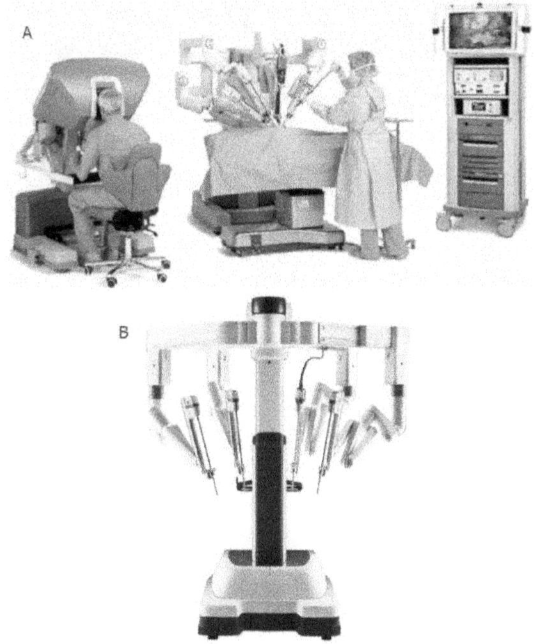

Figure 1.8: A: da Vinci complete surgical robotic systems have three major components: (1) the surgeon's console; (2) a surgical cart with the robotic arms and end-effectors; and (3) the visual cart (copyright Intuitive Surgical International - reproduced courtesy of the manufacturer); B: the latest iteration of the surgical cart for the da Vinci Xi Surgical System (da Vinci, Intuitive Surgical, Sunnyvale, CA, USA)

Source: (Gumbs et al., 2020)

AI-powered robotic systems are also improving surgical planning and decision-making. Machine learning algorithms can analyze patient scans, medical history, and similar cases to recommend the best surgical approach. In orthopedic surgery, for example, AI models can predict the optimal positioning of implants and suggest adjustments during the procedure to improve long-term outcomes. Automation in surgery also extends to intraoperative monitoring, where AI systems continuously analyze vital signs and adjust surgical parameters to maintain patient stability. Furthermore, autonomous robotic systems are being developed for specific tasks such as suturing, tissue retraction, and wound closure, reducing the workload on surgeons and improving surgical efficiency. By combining robotics with AI and automation, modern surgery is becoming safer, more accurate, and more effective. These technological advancements are not only enhancing the capabilities of surgeons but also improving patient care and clinical outcomes, marking a new era in surgical innovation.

1.4.1 AI-Enhanced Minimally Invasive Surgery

Minimally invasive surgery (MIS) is now very dramatically facilitated by powerful software enabled the use of robots that has made the surgery more precise, with quicker recover, much less risks for the patient. MIS involves small incisions performed through specialized tools and cameras with a less traumatic effect on the body from traditional open surgery. Real time guidance through AI powered systems, accurate imaging with better control over the instrument are the revolutionized of MIS by AI powered systems. All the preoperative imaging data including the CT scans, MRIs or fossils are processed by machine learning algorithms to get detailed 3D models of the surgical site. The models provide surgeons with a means to plan the highest likely surgical approach and anticipate potential complications. AI systems present heightened visualization of real-time data poured in from the cameras and sensors during surgery to integrate this into what the robot does, improving precision down to the levels of sub-millimeters. This reduces the likelihood of accidental tissue damage and makes it easier to perform the procedure with the most precision possible. Real time anomaly detection is also enabled by AI to enable the system in being able to flag unexpected changes in tissue structure or bleeding, so that the surgeon can acted upon this quickly to avoid complications.

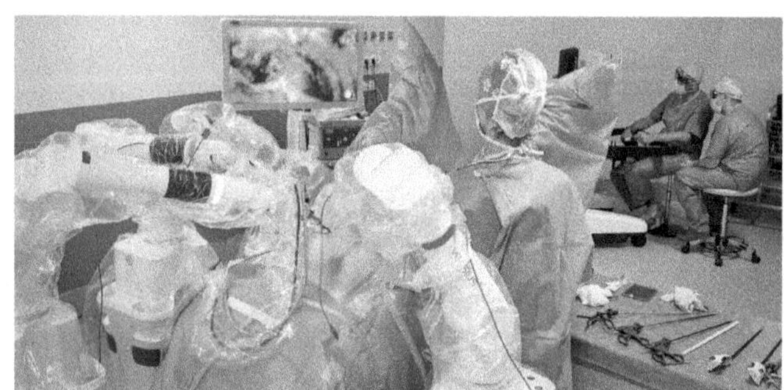

Figure 1.9: The Versus Surgical Robot.

Source: (Halabi et al., 2024)

Real time feedback and improved accuracy of tissue identification in MIS are being afforded to intraoperative decision making by AI. AI image recognition of blood vessels, nerves and tumor helps to avoid surgical errors and increases its accuracy. To take the tip massaging field in media I work in a bit further, I was introduced to AI enhanced fluorescence imaging, which allows health tissue to be distinguished from cancerous tissue, more accurate removing of tumor. Machine learning models look at surgical video feeds to patient vitals to predict problems, and suggest interventions during the procedure, with the aim of anticipating poor prognoses. Such autonomous and semi-autonomous robotic systems powered by AI can assist in the complex tasks like suturing, tissue retraction and cauterization that enables the surgeon to concentrate on the most critical part during the surgery. Furthermore, AI can forecast the best path for surgical instruments to take, decreasing the probability of harm to on the spot tissues and enhancing total surgical outcomes general.

Other means of MIS improvement, such as AI-based virtual and augmented reality (VR/AR) systems, that allow surgeons to see how a patient's tissue will appear in 3D once viewed in context to adjacent tissue and organs are becoming increasingly better. AR overlays the ongoing data and 3D models of the surgery to the surgical field, helping the surgeons to see the anatomical structures and surgical instruments better. That way, navigation and positioning can be more precise in the case of more complex procedures. These visualizations are then improved by AI algorithms that predict the movement of organ or tissue in response to surgical manipulation such that surgeons can adjust during the procedure.

AI has also been applied to MIS to benefit postoperative care and recovery. Using AI models, patient data (vital signs, mobility, pain levels) can be analyzed to be able to predict on the possibility of complication and early intervention. Patients at risk of infections, or delayed recovery can be predicted by the models and corrective care plans can be adapted. Robotic systems that are driven by AI can also help people in rehabilitation, guiding the patient through personalized recovery exercises following the state of their progress and limitations.

However, integrating AI into minimally invasive surgery provides hospital an additional edge of improved patient safety, shorter recovery times, and higher surgical success rates overall. Beyond one's failure prediction, AI not only improves the accuracy of the surgical procedures, but also helps improve patient outcomes, speed recovery, as well as reduce postoperative complications. This technological advance is in each way not just changing the picture postoperative, but additionally boosting the ubication of hurriedly complex projects, making them safer and more effective.

1.4.2 AI in Real-Time Surgical Decision Support

Real time surgical decisions support is made by AI to help surgeons make immediate insights, predictive analysis, and action recommendations during the procedures. Traditional surgical decision-making is human cognitive, which has limits in perception and cognitive load. This process is made superior with the help of AI that takes real time data from imaging systems, surgical instrument and patient vitals to offer intelligent guidance and also warn about possible complications at an early stage. The huge intraoperative data volume is processed by machine learning algorithms to draw attention to patterns and anomalies which a human eye may fail to notice. Such is the case for AI, which can detect small differences in the texture of tissue or blood flow, and if somehow a surgeon can adjust his or her approach before issues begin to occur. AI based predictive models can predict, for example, edge of the cord injury or excessive bleeding and immediately suggest corrective measures to improve patient safety and the rate of success of the surgical procedure (Vannaprathip et al., 2022).

Real time feedback is given in the form of AI driven systems that increase surgical precision and consistency by increasing the feedback for extension of multiple instruments used during surgery along with incision depth. Endoscopic and laparoscopic surgeons use the live video feed from endoscopic or laparoscopic cameras for instance computer vision and deep learning model to distinguish different tissue types and anatomical structures with a greater accuracy. For instance, AI can help blood vessels, nerves or tumors shown in real time so as not to cause accidental damage. Haptic feedback systems augmented by AI can also instruct the surgeon's hand movements away from harm through resistance changes of the system and recommendations for angles of cutting or stitching. Furthermore, AI can use data from similar past procedures to suggest the most appropriate surgical method according to the patient, and so on. In addition to making the best surgical outcomes, this also improves the duration of surgical procedures and recovery times.

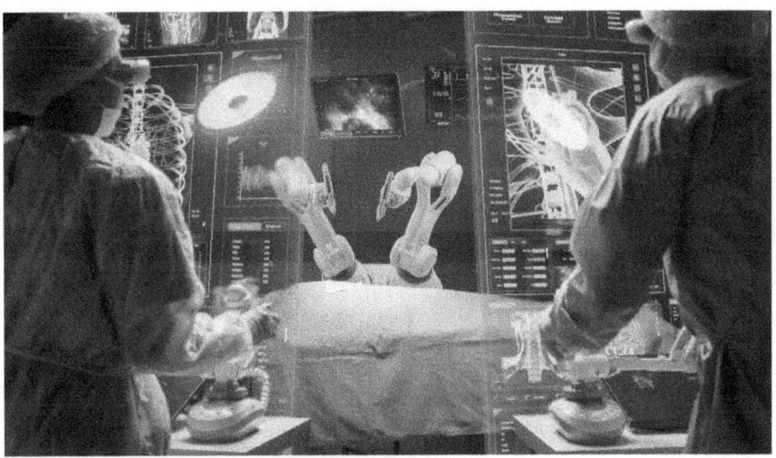

Figure 1.10: Experts discussed the intersection of surgery and artificial intelligence at the ACS Clinical Congress.

Source: (Mohamadipanah et al., 2022)

It is also helping robotic assisted surgery by making it more automated and flexible. AI helps with robotic surgical platforms like the da Vinci to adjust the movement in real time dynamically as the tissue response together with the surgical progress takes place. Within surgical tools, there are AI systems that are constantly measuring the position, and how pressure is applied to surgical tools, and compensating for tremor or misalignment automatically so as to keep as much as precision as possible. AI based augmented reality (AR) overlay can also give surgeons more visual guidance like real time 3D map of the organs as well as predicted tissue movement. By this they help the surgeons navigate through complex anatomical structures better and allow them to adapt their approach dynamically. Using AI to combine real time data analysis and analysis, predictive modeling and robotic assistance, AI is changing surgical decision making resulting in safer, more accurate and efficient procedures.

1.5 Personalized Medicine and Genomics

AI is revolutionizing the personalized medicine and genomics space by enabling the development of personalized therapies on someone's genetic profile, their medical history as well as lifestyle. The difference between traditional medicine and AI medicine is that the traditional medicine does not perform much of the data, so most of the time they use a one size 'all' approach. But AI medicine can be very specific with some tailored treatments by looking at the whole genomic and clinical data. Alt. Complex genetic information can be processed for identification of disease risk factors and personalized treatments with such machine learning and deep learning algorithms as a DNA sequences, gene expression patterns, and protein interactions. For instance, AI can forecast a patient's likelihood of getting such diseases as cancer or cardiovascular conditions and recommend primary prevention, early intervention or behavioral measures. AI is also being employed in an enhancement of Pharmacogenomics, which assesses the effects of an individual's genes on their readings to drugs. For example, genetic markers are one thing AI can be used to analyze to predict how a patient will metabolize medications and then change the drug dosages and select the best treatments and plan that is least likely to cause side effects(Patrinos & Mitropoulou, 2022).

Similarly, AI is helping in the diagnosis and treatment or genetic disorders by identifying disease causing mutations and suggesting appropriate therapies. AI algorithms are being used in oncology to see how genetic mutations in cancer cell determine the accuracy of a precision treatment. An example is that AI can identify particular mutations in oncogenes and tumor suppressor genes, thus assisting oncologists in choosing therapies targeted at those mutations. Much like diabetes mellitus can reach its diagnosis and treatment stage much faster for many patients, AI is speeding up the identification of novel biomarkers—molecular markers of disease—but from large genomic datasets. The biomarkers provide more precise and earlier diagnosis of disease, which is important for the patients. At the same time, AI is helping researches in gene editing technologies like CRISPR to identify a target sequence and predict a potential off target effect, thus enhancing accuracy and safety of gene editing. Rapidly, AI models can find previously unknown mutations in rare genetic diseases and help in granting gene targeted therapies to shorten the time to diagnosis and to give better treatment options.

Moreover, AI is assisting in personalized medicine through calculation of polygenic risk scores, which are the result of calculation of genetic risk to develop complex diseases such as diabetes, hypertension and Alzheimer's. People's risk profile can be provided via AI models using data from genome-wide association studies (GWAS) and other sources, and told that they should take appropriate preventive actions. Risk patients can be Intervened early by means of high-risk patients via predictive models such as Identification through lifestyle changes, medication, or increased monitoring. Along with that, the clinicians are using AI driven tools for personalized rehabilitation and recovery programs, which are customized according to the genetic and physiological profile of a patient. As AI advances, personalized medicine and genomics are sparking a foray for AI toward changing health care by making it more precise, increasing diagnosis capability, and delivering more effective treatments with a better patient care.

1.5.1 AI for Genetic Risk Prediction

AI is playing a transformative role in genetic risk prediction by analyzing complex genetic data to assess an individual's predisposition to various diseases. Traditional methods of predicting genetic risk have relied on identifying single gene mutations or family history, which often provide limited insights into complex diseases. AI, particularly machine learning and deep learning algorithms, can analyze vast amounts of genomic data, including single nucleotide polymorphisms (SNPs), gene expression profiles, and epigenetic markers, to identify patterns and correlations that may not be apparent through conventional analysis. Polygenic risk scores (PRS), which combine the effects of multiple genetic variants to estimate the likelihood of developing complex diseases, are being enhanced by AI models that can process large-scale genome-wide association studies (GWAS) data. AI-based models can integrate genetic information with environmental and lifestyle factors to provide more accurate and personalized risk assessments. For example, AI can predict an individual's risk of developing diseases such as diabetes, cancer, and cardiovascular conditions, allowing for early preventive measures and targeted interventions.

AI is improving the accuracy and scope of genetic risk prediction through advanced pattern recognition and predictive modeling. Deep learning models can uncover complex interactions between genes and environmental factors, providing a more comprehensive understanding of disease susceptibility. For instance, AI algorithms have been used to predict the likelihood of developing breast cancer based on genetic markers and mammogram data, achieving higher accuracy than traditional statistical models. In cardiovascular disease, AI-driven models can analyze genomic data along with clinical indicators such as cholesterol levels and blood pressure to predict heart disease risk with greater precision. AI systems can also integrate real-time health data, such as wearable device information and electronic health records (EHR), to adjust risk profiles dynamically. This allows for more responsive health management, where early warning signs can trigger tailored preventive strategies, such as medication adjustments, dietary changes, or increased monitoring.

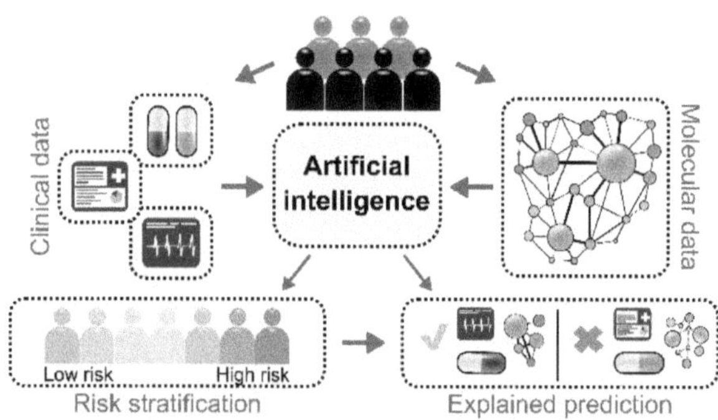

Figure 1.11: AI use in medicine to stratify risk and explain predictions using clinical and molecular data.

Source: (Westerlund et al., 2021)

AI's ability to handle large-scale genetic data is accelerating the identification of previously unknown genetic risk factors. Machine learning models can analyze rare mutations and identify subtle genetic variations associated with increased disease risk. For example, AI has been used to identify rare mutations linked to neurodegenerative disorders, enabling earlier diagnosis and more targeted treatment options. AI models can also predict how combinations of genetic variants contribute to disease susceptibility, which is particularly valuable for complex diseases like Alzheimer's and autoimmune disorders. Additionally, AI is helping to identify gene-environment interactions that influence disease progression, such as how dietary habits, pollution exposure, and stress levels affect genetic expression. This deeper understanding enables more precise and individualized health interventions, improving patient outcomes and reducing long-term health risks.

Furthermore, AI is facilitating population-level studies, allowing researchers to understand how genetic risk varies across different ethnic and demographic groups. Traditional genetic studies have often been biased toward specific populations, leading to gaps in understanding genetic risks for underrepresented groups. AI-driven models can analyze diverse genomic datasets to develop more inclusive and accurate risk prediction models. This helps ensure that genetic risk assessments are applicable across different populations and healthcare systems. By providing more precise and comprehensive genetic risk assessments, AI is enabling a shift toward proactive and preventive healthcare, where interventions are tailored to an individual's genetic and environmental profile. This approach not only improves patient outcomes but also reduces the overall burden on healthcare systems by preventing disease progression and minimizing the need for costly treatments.

1.5.2 AI-Driven Personalized Treatment Plans

Personalized treatment plans are revolutionized by AI that helps healthcare providers to design the most personalized and individualized medical interventions built on patient's genetic configuration, medical history, lifestyle, the environment. Traditional treatments and traditionally used topotecan follow generalized protocols that have not considered individual aspects and thus result in variable outcomes and potential side effects. Databases of electronic health records (EHR), proteins, genomic sequences, imaging techniques, as well as reports from patients can be analyzed by AI powered systems such as machine learning and deep learning models to identify common piece patterns and correlations that a better treatment result. A combined genetic and clinical data can be integrated using AI and could recommend targeted therapy that will be specifically tailored to the patient's biological and physiological characteristics. For instance, AI models may use genetic profile of a tumor to suggest the most successful chemotherapy or immunotherapy and thereby enhancing the treatment response and lowering toxicity.

However, personalized treatment plans that are driven by AI work excellently in managing the chronic diseases and complex medical conditions. For example, AI models can analyze a patient's glucose levels, dietary habits, and physical activity to suggest personalized insulin dosages and meal plans, these models can also evaluate the supplies required by a patient during hospital stay. Just like that, AI based platforms are being used to design the personalized medication regimens for patients with hypertension and cardiovascular diseases taking into account blood pressure patterns, genetic markers and patient's medication response history. Additionally, AI can forecast how a patient will assimilate various drugs depending on his or her genetic data, which can then influence doctors in adjusting dosages and mixtures of drugs for the best results and least side effects. This pharmacogenomic approach diminishes drug toxicity risk, thereby making treatments safer by ensuring that they will be more effective so that patients will be better compliant.

Dynamic adjustments to treatment plan are made according to real time data and AI is capable to analyze it. For instance, in an intensive care unit (ICU), AI algorithms observe the vitals of the patient, and lab results, constantly and when complications start, such as sepsis or organ failure, they are detected first. Immediately on recognizing the deterioration, the AI system can then recommend what the shortcuts should be for now to prevent deterioration — like fluid management or antibiotic administration. AI models can also monitor tumor progression during treatment in oncology and advise how to adjust them by changing drug combinations or even introduce targeted therapies, based on genetic mutations of tumor cells, which appear during treatment. Newer of AI driven platforms are also taking care of patient's post-surgical recovery by analyzing his patient reported outcomes and biometrics from wearable devices to adjust his rehabilitation programmers and pain management strategies in real time.

AI is also helping precision medicine in the sense that it is combining all types of data sources for a comprehensive patient profile. Genomic data can be combined with that of proteomics (protein analysis), metabolomics (metabolic profiling) and that of the microbiome to yield an entire picture of a patient's health status using AI models. It allows for a more accurate disease modeling and predictive analytics to help clinicians predict a course of disease and determine appropriate preemptive treatment plans. There are also AI driven virtual care platform that can engage patients directly with personalized health recommendation, reminders of medications and lifestyle adjustments through mobile apps and digital health tools. AI introduces data-driven treatment strategies with highly customized strategies, therefore in delivering better patient care, enhancing clinical outcomes, improving resource utilization and targeting interventions that help reduce the healthcare costs.

1.5.3 AI in Rare Disease Diagnosis

A growing number of rare diseases, may be complex and poorly understood medical illnesses, are being more quickly and accurately diagnosed using AI. Developing a rare disease is notoriously difficult: There are rare diseases that affect a very small proportion of the population, have relatively overlapping symptoms with more common diseases, and rare medical knowledge. The traditional diagnostic methods are generally lengthy processes and a patient may need to visit several specialists and undergo numerous tests, to reach the final diagnosis. Machine learning and deep learning models with their ability to analyses large medical datasets such as genetic information, clinical records and images can accelerate this process by analyzing huge number of medical datasets that may go unnoticed by human clinicians for rare patterns of diseases. By comparing a patient's symptoms to stored disease profiles and medical literature in AI models, there is an increased chance of accurate diagnosis at an earlier stage than is possible by comparison. For instance, AI based systems have been employed in diagnosing rare genetic disorders like Ehlers-Danlos syndrome, Marfa syndrome, etc. by genetic mutations and symptom groups better than conventional diagnostic tools.

AI has a major impact in genomic analysis to diagnose rare disease. Most develop from the misuse of genes. Therefore, AI based algorithms can process whole genome and exome sequencing to locate disease causing mutations. Because the rare disease speed is dictated by the nature of the material to be regrouped (single nucleotide polymorphisms (SNPs), copy number variations (CNVs), and structural variations that play a role in the rare genetic condition) Using this approach, machine learning models have become able to detect them. In addition, AI can help identifying novel gene disease associations by analyzing large genomic data from already known biological pathways as well as the phenotypic traits. For example, AI has been used to find new mutations related to neurodevelopmental or metabolic diseases to more precisely diagnose and treat the suspected conditions. In addition, genetic data and clinical or environmental factors can be combined by use of AI based platforms to offer a more complete understanding of the patient's condition to aid differentiating between rare diseases that show similar signs.

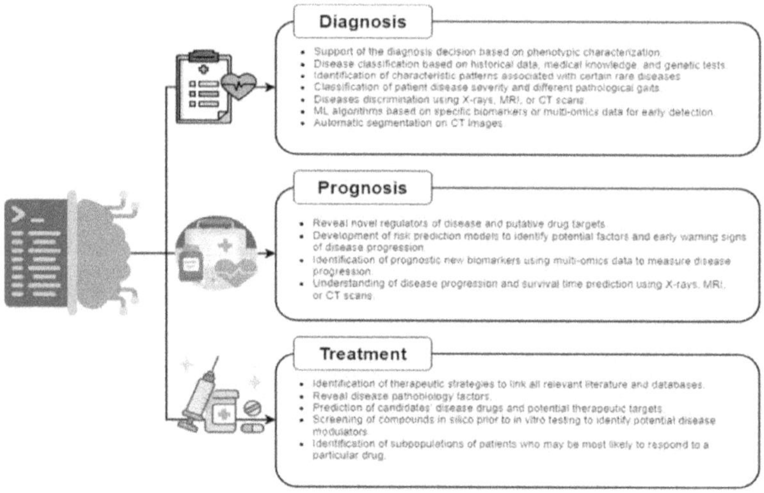

Figure 1.12: Examples of ML applications within diagnosis, prognosis, and treatment.

Source: (Visibelli et al., 2023)

AI was also used to improve imaging analysis for diagnosis of rare diseases. Radiological images such as MRI or even CT scans can be analyzed by deep learning models to detect rare neurological, skeletal or cardiovascular abnormality that may be feature of rare diseases. For instance, AI has been failing to find patterns of atrophy of the brain linked to rare neurodegenerative diseases like Huntington's disease and amyotrophic lateral sclerosis (ALS). Also, the retinal scans can be analyzed by AI algorithms to determine if there's a rare eye disease or a genetic syndrome with ocular manifestations. In other cases, AI could catch some very fine anomalies that are difficult to see by the human eye, thus diagnosis will be early and accurate. Some AI tools are also being used to identify genetic syndromes defined by the facial structure, including Noonan and Cornelia de Lange syndromes. It… helps around diagnosis speed and accuracy by pulling in geometrical information along with genetic and clinical information and making AI more precise at diagnosing patients with rare conditions.

Furthermore, AI enhances the creation of global rare disease databases and diagnostic platforms that can better facilitate sharing of knowledge and working with medical professionals and researchers. AI driven platforms allow multiple data sources such as electronic health records (EHR), research studies as well as patient registry to aggregate patient data to construct complete disease profile. These data can be used by machine learning models to learn to detect types of diseases that have gone unrecognized before and also improve the diagnostic guidelines. Real time diagnostic recommendation and risk assessment is also facilitated using the AI system that is based on patient genetic and clinical profile to support clinical decision making. For rare disease diagnosis, as AI keeps advancing, it is expected to improve the diagnostic delays and increase the diagnosis accuracy and allow the discovery of new targets for the treatment. Data driven, AI assisted diagnostics are both a possibility and pose to assist patients with rare diseases with better outcomes and to ease the burden on healthcare systems in general.

Multiple Choice Questions (MCQs)

1. **What is one major way AI is improving diagnostic accuracy in medicine?**

   a. Increasing the number of available doctors

   b. Enhancing the expertise of radiologists

   c. Analyzing medical images and detecting subtle anomalies

   d. Increasing the cost of medical procedures

2. **How does AI enhance treatment planning for patients?**

   a. By replacing doctors entirely

   b. By analyzing genetic data and patient response patterns

   c. By limiting the number of treatment options

   d. By providing the same treatment to all patients

3. **Which imaging technique has AI improved for detecting neurological disorders?**

   a. Ultrasound

   b. Magnetic Resonance Imaging (MRI)

   c. X-ray

   d. CT Scan

4. **What advantage do AI-powered wearable devices provide in patient monitoring?**

   a. They eliminate the need for hospital visits

   b. They predict emergencies and improve response times

   c. They only monitor heart rate

   d. They work only when connected to a computer

5. **How does AI contribute to drug discovery and development?**

   a. By performing drug trials without human involvement

   b. By analyzing large biological and chemical data to identify promising candidates

   c. By developing new chemical formulas independently

   d. By increasing drug manufacturing costs

6. **How does AI enhance surgical precision and patient outcomes?**

   a. By reducing the need for surgeons

   b. By suggesting optimal surgical techniques and providing real-time feedback

   c. By automating the entire surgical process

   d. By increasing tissue damage during surgery

7. **What role does AI play in personalized medicine and genomics?**

   a. It identifies common treatments for all patients

   b. It creates identical treatment plans for every individual

   c. It enables tailored therapies based on genetic profile and lifestyle

   d. It eliminates the need for human doctors

8. **What type of genetic analysis does AI use to predict disease risk?**

   a. Clinical trials

   b. Polygenic risk scores

   c. Radiology reports

   d. Heart rate monitoring

9. **How is AI improving clinical trials?**

    a. By reducing the number of participants

    b. By predicting outcomes and improving patient recruitment

    c. By relying solely on human observation

    d. By increasing the time required to complete trials

10. **In oncology, how does AI help identify targeted therapies?**

    a. By providing the same treatment for all patients

    b. By detecting specific mutations in oncogenes and tumor suppressor genes

    c. By analyzing blood pressure alone

    d. By delaying the diagnostic process

Answers:

1	2	3	4	5	6	7	8	9	10
c	b	b	b	b	b	c	b	b	b

AI in Healthcare Operations and Management

2.1 Optimizing Hospital Administration and Workflow

A hospital administration is the handling of the overall operations and available resources of a healthcare facility to deliver the patient care in an efficient and effective way. Strategic planning, financial management, human resource allocation, and compliance with health care regulations are included in it. Policies are set, patient care standards improved, budgets are managed to ensure smooth operations for the hospital by administrators. They strive to provide a better healthcare service delivery by ensuring staff works properly, ensures adequate utilization of resources and developing culture that is patient centered. Also, effective administration entails working with medical staff, technology providers, and insurance companies to continue the uninterrupted system. To respond to these challenges, such as staff shortage, increasing operation costs and the development of healthcare technology, administrators have to use data-driven decision making and adopt modern healthcare management systems. For the effective hospital administration, it is essential to have strong leadership, transparent communication, and hospital performance to be reviewed constantly(Desarno et al., 2021).

Hospital workflow is the order of the processes and tasks in hospital care and operations. This is also patient admission, diagnosis, treatment, discharge, and follow up care. Workflow optimization is about diminishing bottlenecks, cutting down patients waiting times and better orchestrating the departments within the radiology, pharmacy, labs... This forms of automation and digitalization including the use of electronic health records (EHR) and automated scheduling system improves workflow efficiency. For instance, it (streamlining patient triage and automating diagnostic result processing) helps reduce delays and facilitate the making of a decision. Healthcare providers, nurses and support staff together can communicate effectively and help patients get timely care. Proceeding to the implementation of predictive analytics and artificial intelligence, helps hospitals forecast patient inflow and predict resource demand, which can further improve workflow by predicting patient arrival and resource demand with a view to manage staff and equipment efficiently. A good workflow is optimized and better patient outcomes, lower operational costs and better performance of the entire hospital.

2.1.1 AI for Patient Scheduling and Resource Allocation

The healthcare operations are being transformed by artificial intelligence (AI), which is improving in patient scheduling and resource allocation. To improve patient care, reduce the costs of operations, and improve hospital overall efficiency, accurate patient scheduling and resource management is very vital. However, traditional scheduling methods tend to also experience problems like long patient wait time, inefficient use of medical resources as well as staff burnout resulting from patient inflow unpredictability. With AI, these challenges have a data driven answer that look into historical data, model patient demand and come up with them with ideal schedules that balance between patient requirements and available sources(Shaw, 2023).

AI in Patient Scheduling

Patient scheduling systems based on AI combined with machine learning algorithms and predictive analytic provide better accuracy and more efficiency of appointment getting and patient flow management.

Figure 2.1: Appointment scheduling software.

Source: (Niu et al., 2023)

These systems analyze historical patient data with respect to appointment, no show rate and seasonal variation of patient volume to predict the future demand and then fix the schedules. For instance, AI can determine the high turnout and automatically ... when we ... more staff and resource in order to serve more numbers of patient flow, thus raising the patient satisfaction and lowering the patient waiting length.

- **Predictive Scheduling:** Historical data and real time input can be used to predict a patient's consumption predictions using AI systems and they can be used by the hospitals to predict volume fluctuations in patient demand. It means that the right amount of staff and medical resources are always available during busy hours and not wasted during slack hours. Predictive scheduling minimizes the wait time and improves patient satisfaction by matching the patients' needs with available resources.

- **Automated Appointment Management:** Natural language processing (NLP) and chatbot based systems go on to make patient scheduling even easier for patients as they are able to do so through automated platforms and book, modify or cancel appointments. AI driven systems can look at symptoms and urgency of the patients and schedule appointments according to the severity of the condition. This means that serious cases get their turn and those that aren't so serious won't overwhelm the hospital staff.

- **Integration with Electronic Health Records (EHR):** AI-based scheduler systems work to connect with electronic health records (EHR), meaning an appointment's data will really be entered in real time.

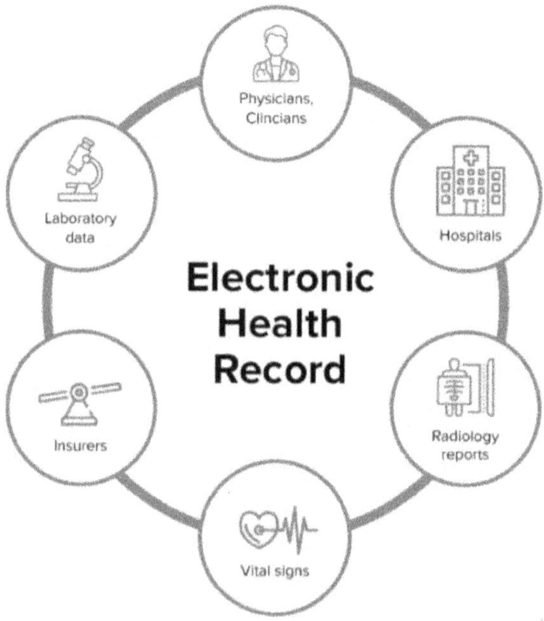

Figure 2.2: Electronic Health Record.

Source: (Evans, 2016)

It helps to eliminate scheduling conflicts and eliminates double booking of resources such as diagnostic equipment and consultation rooms. Also, AI can also send automated reminders to patients through text message or email and thereby reducing high rate of missed appointments and improving the overall scheduling efficiency.

AI in Resource Allocation

The predictive modeling and real time data analysis that AI provides benefits resource allocation by helping the hospital resources, staff, equipment and facilities to work at optimal levels. The resource demands can be predicted by machine learning models that analyze the patient admission trends, treatment durations and the discharge patterns. It allows hospital administrators to set the correct levels of staffing, to distribute medical equipment and to organize hospital bed.

- **Bed and Facility Management:** Patient admissions can be monitored in real time and hospital beds predicted to become available on the basis of how long patients will remain in bed and when they will be discharged. It prevents overcrowding and allows people to staff and resource critical care units and emergency rooms adequately. AI can also schedule and minimize downtime, analyzing the availability and use of diagnostic equipment such as MRI and CT scanners. AI helps predict when equipment will be needed for maintenance and repair, allowing them to be there at due times, thus preventing delay in patient care.

- **Staff Scheduling and Workload Balancing:** AI as well can use to improve the process of staff management by observing patterns of workload and offering balanced shift schedules. It minimizes staff fatigue and burnout and thus, improves patient care and increases the job satisfaction of healthcare providers. For example, AI can determine whether to adjust employee working hours according to patient demand or the employees' available time.

- **Inventory and Supply Chain Management:** Supply chain management in hospitals is supported by AI which predicts demand for medications, surgical supplies and other consumables. With the help of usage patterns as well as external factors including supply chain disruptions, AI will be able to suggest inventory levels which avoid shortages and waste. AI can also power automated procurement systems that automate the ordering process, so items that are essential are available whenever they are needed without excessive stocking. Hospitals can increase operation efficiency, improve patient care quality, and provide better patient experience through the exploitation of the AI capability in scheduling of patients and resource allocation. AI provides value for modern healthcare management because it is capable of processing large volumes of data, as well as finding patterns, and making real time decisions.

2.1.2 Reducing Administrative Burden with Automation

Businesses definitely do need administrative tasks, but they occupy a large percentage of time and resources. Some of these tasks include data entry, scheduling, document processing, invoicing, compliance tracking, internal communications. Although their significance, administrative work is largely repetitive and demanding upon time, taking much administrative effort away from more strategic and value creating work. Artificial intelligence (AI) and machine learning (ML) powered automation has successfully arrived to take on the administrative burden, improve efficiency, and increase productivity, to a large extent(Kaplan & Anderson, 2003)the use of subjective and costly-to-validate time allocations, and the difficulty of maintaining and updating the model as (i.

Use of automation reduces the time for doing routine or rule-based jobs without human involvement. With the help of AI driven automation tools, such as data extraction tools, invoice processing tools, the schedule meeting tools, the email communication tools and the chatbot for customer service inquiries. For example, natural language processing (NLP) lets an AI agent answer to customer emails so that human input is not required. Also, Machine Learning Algorithms can categorize documents in a machine way to file these documents in such a manner that they would contain accurate and organized data.

The biggest advantage of automation is related to the decrease of human error. Manually handled tasks in administrative duties are liable to errors because of fatigue, oversight, and data incoherence. When such automatic systems are used, risks are decreased because the information is always processed the same way and precisely. For instance, an AI-based invoice processing AI can process invoices automatically extract the details, check the details against purchase orders and denote discrepancies which minimizes payment errors and increases financial accuracy.

It also allows better allocation of resources. With the help of AI systems for routine tasks driven off the human employees can dedicate their time into high value activities like strategic decision making, customer engagement and development of business. This transition contributes to job satisfaction and boosting of labor force efficiency as employees are no longer lost in tedious low value jobs. Additionally, automation systems are 24/7 operational meaning the administrative processes will not stop, allowing for faster turnaround times and faster response time for businesses.

Figure 2.3: Simplifying Billing with Automated Processes

Source: (Iqbal, 2025)

Automation on the other hand not only brings about internal benefits but also alleviates administrative burden thus improving customer service experience. For instance, automated customer service systems, can give off-the-cuff answers to common queries, and send complex ones to humans. Through the coordination with the customers and staff in real time, the automated appointment scheduling systems can reduce the scheduling conflicts and increase the efficiency of service.

Additionally, AI powered automation is supportive for a compliance and reporting. Documentation and reporting are often rigorous and time-critical, both of which may be automation-able for accuracy and consistency in dealing with regulatory requirements. Such automated systems can automatically produce compliance reports, contain the key performance indicators (KPIs), and notify managers when they are potential problems have been identified — assisting organizations in remaining compliant while minimizing the manual labor.

Overall, automated administrative is reduced administrative burden and, as a result, increases operational efficiency, decreases cost, improves accuracy and increases employee and customer satisfaction. With advancements in AI and automation technologies, businesses will be able to take advantage by free up human resources to more strategic and creative roles where businesses can be unconquerable as compared to man.

2.2 Predictive Analytics for Patient Care

Healthcare providers are starting to predict the health outcomes of patients and therefore can decide who is at risk, and what proactive actions to take to prevent complications. Predictive models can find patterns and predict the future health events when analyzed big datasets, which may include electronic health records (EHRs), medical imaging, lab results et cetera. Machine learning can help to identify early signs of such chronic diseases such as diabetes, heart disease and cancer that may be able to make early intervention and improve patient outcomes. AI models doing the same using MRI and CT scan data can detect tumor growth far earlier than currently offers diagnosis and target the cancer more effectively. It also helps hospital management predict patient ad missions, discharge rates, and resource requirements thereby enhancing better staffing, optimal bed availability and efficient inventory management. In addition, predictive models may also help healthcare providers predict the likelihood of readmission due to a certain factor like patient age or an underlying condition and initiate more targeted follow up care and decrease unnecessary trips to hospitals.

In addition to being able to diagnose diseases more quickly than at other times, and for real time patient monitoring: predictive analytics also helps in designing personalized medicine. Pharmacogenomics is defined as the combining of genomics and pharmacology to develop prediction models that permit determining the impact of the patient's genetic makeup on his or her response to medications. This narrows down which drugs and dosages are prescribed by healthcare providers and the least likely to cause side effects. Wearable devices, remote monitoring systems as well as its real time tracking of the vital signs, activity, sleep patterns further strengthen the predictive analytics. This data analysis can be done by AI models to identify early signs of health deterioration and a timely intervention can reduce medical emergencies. On a larger scale, predictive analytics allows for population health management by spotting the disease trends and risk factors of common groups of people, and helps with making improvements on a larger scale. Thus, it makes it easier for healthcare systems to allocate resources more efficiently, optimize public health initiative and adopt effective preventive care strategies. Finally, predictive analytics confronts healthcare as a reactive rather than proactive system and improves patients' outcomes, lowers healthcare costs, and overall care quality.

2.2.1 AI in Early Disease Prediction and Prevention

AI is playing a pivotal role in early disease detection, significantly enhancing diagnosis and preventive care with the help of ambulances. By analyzing various data sources such as genetic information, lifestyle factors, and medical history, AI algorithms can identify patterns and risk factors associated with certain conditions. Zeitzer says AI algorithms can be really helpful in the healthcare industry; it helps in identifying high-risk individuals and implementing targeted preventive measures. By detecting diseases at an early stage, healthcare providers can intervene proactively and significantly improve patient outcomes(Scholz et al., 2022)regarding correlations between stroke detection at the EMS and stroke specific, as well as personal characteristics such as stroke type, sex, age, weekday, time of day, year, EMS number contacted, and treatment. The possible increase in stroke detection through an ASR and the effect on stroke treatment was calculated based on the impact of an existing ASR to detect OHCA from CORTI AI. Results: The Chi-Square test with the respective post-hoc test identified a negative correlation between stroke detection and females, the 1813-Medical Helpline, as well as weekends, and a positive correlation between stroke detection and treatment and thrombolysis. While the association analysis showed a moderate correlation between stroke detection and treatment the correlation to the other treatment options was weak or very weak. A potential increase in stroke detection to 61.19% with an ASR and hence an increase of thrombolysis by 5% in stroke patients calling within time-to-treatment was predicted. Conclusions: An ASR can potentially improve stroke recognition by EMDs and subsequent stroke treatment at the EMS Copenhagen. Based on the analysis results improvement of stroke recognition is particularly relevant for females, younger stroke patients, calls received through the 1813-Medical Helpline, and on weekends. Trial registration: This study was registered at the Danish Data Protection Agency (PVH-2014-002.

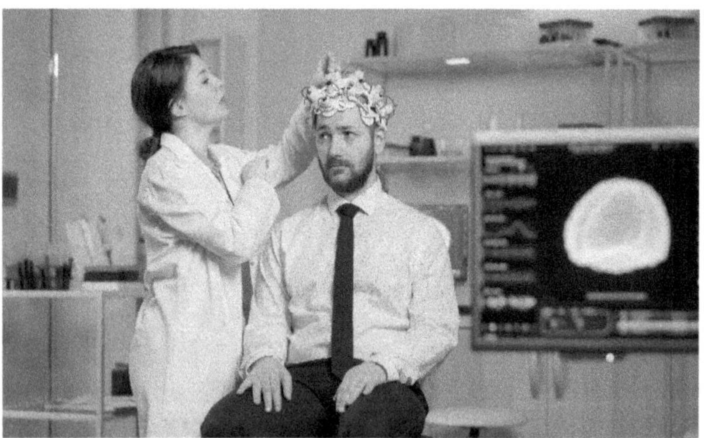

Figure 2.4: Preparing for Brain Activity Measurement.

Source: (Alivisatos et al., 2013)

Sweta Mangal Zeitzer knows the importance of Machine learning, and she knows that machine learning is a subset of AI, and plays a vital role in preventive medicine. By analyzing large datasets, machine learning algorithms can identify hidden patterns and correlations, allowing healthcare professionals to make accurate predictions about disease development and progression. Healthcare organizations have been leveraging machine learning techniques to develop preventive medicine strategies. By identifying individuals who are at a high risk of developing certain diseases, healthcare providers can offer personalized interventions, such as lifestyle modifications or targeted screenings, to prevent or delay the onset of the condition.

- **AI-enhanced screening and Early Intervention Programs**

  Early detection of diseases is crucial for successful treatment and improved patient outcomes. AI has enhanced screening programs by improving accuracy and efficiency. Through the analysis of medical images, such as X-rays, MRIs, and CT scans, AI algorithms can identify subtle anomalies and early signs of diseases like cancer, cardiovascular conditions, and neurological disorders. Healthcare organizations, including Zeitzer Healthcare Ltd, says, AI-enhanced screening programs help to identify patterns and deviations that may indicate the presence of underlying health conditions. This early detection allows for prompt intervention and targeted treatment, potentially preventing the progression of diseases and improving patient outcomes. AI in screening programs highlights their commitment to utilizing innovative technologies for the benefit of patients' health.

- **Integrating Wearable Devices and AI for Continuous Monitoring**

  Wearable devices, such as fitness trackers and smartwatches, have gained popularity in recent years. These devices can collect real-time health data, including heart rate, sleep patterns, and activity levels. By integrating wearable devices with AI algorithms, healthcare professionals can continuously monitor individuals' health and detect deviations from normal patterns. Zeitzer Rajasthan knows the value of integrating wearable devices with AI technology to collect real-time health data and provide continuous monitoring. The Ai integration allows for early detection of health issues, timely interventions, and improved preventive care.

2.2.2 AI for Hospital Readmission Risk Analysis

Healthcare providers can identify patients at high risk for returning to the hospital within a short period of time of discharge thanks to AI. Hospital readmissions are an expensive part of care for healthcare systems and may indicate gaps in patient care or follow up. The input to AI models involves a wide variety of patient data like electronic health records (EHRs), vital signs, lab results, medical history and social determinants of health to extract from them patterns and to predict the likelihood of readmission. These algorithms can process all of that complexity of age, underlying conditions, medication adherence, and previous hospital visits etc. to come up with a risk score for each patient. For instance, AI models may signal to that patient with a history of heart failure, having poor medication adherence, and little social support is at high risk for readmission. It permits the healthcare providers to develop and implement personalized discharge plan and follow up care plan, for example home health visits or adjusting of medication, to avoid readmission.

The AI driven readmission risk analysis also improves the hospital operations efficiency by allowing the care teams to focus on high-risk patients and by better allocating resources. Continuous monitoring of the patient data by real time AI systems can alert about the patient's condition deteriorating after discharge. In such case, some vital sign like heart rate or blood pressure can be tracked using wearable devices or remote monitoring systems and then AI models can be applied to analyze this data to find out about early warning signs of complications. It is when a risk of readmission is identified, that healthcare providers can intervene quickly and prevent the need for emergency care. Additionally, AI models can identify those systemic issues that are contributing to the high readmission rate, so that the hospital can take more targeted improvement, for example when the coordination of care is not there or the patient was not educated adequately. Thereby, AI reduces patient readmissions, improves patient outcomes, reduces healthcare costs, and improves hospital performance in general.

How is AI helpful in reducing readmission in hospitals?

Figure 2.5: AI in Healthcare: Four Pillars.

Source: (Javaid et al., 2022)

- **Predictive Analytics:** AI algorithms can analyze patient data to identify individuals at high risk of readmission. By examining factors such as demographics, medical history, and treatment plans, AI can help healthcare providers target interventions effectively. Natural Language Processing (NLP) can analyze patient notes and discharge summaries to identify concerns or risks that might lead to readmission, facilitating better follow-up care and communication.

- **Personalized Care Plans:** AI can assist in creating tailored discharge plans that consider individual patient needs, preferences, and social determinants of health, improving adherence to post-discharge care.AI can streamline communication between different care teams, ensuring that all providers are informed about a patient's care plan and needs, reducing gaps in care.

- **Remote Monitoring:** AI-powered wearable devices and telehealth solutions can continuously monitor patients' vital signs and health conditions post-discharge, alerting healthcare providers of potential issues before they require readmission. AI driven Patient Engagement Tools like tab reminders, tele assistants can enhance patient engagement by reminding them about medications, follow-up appointments, and self-care practices, improving adherence to treatment plans.

- **Effective Resource Allocation:** AI can optimize staffing and resource allocation in hospitals, ensuring that high-risk patients receive adequate support and monitoring during their stay, so that there will be no necessity for further readmissions. AI also can recommend appropriate interventions with real-time decision support tools that help identify potential complications, thus reducing the likelihood of readmission. By leveraging these AI capabilities, healthcare systems can enhance patient outcomes, improve care transitions, and ultimately reduce the rates of hospital readmissions.

2.2.3 AI in Chronic Disease Management

Chronic diseases (such as diabetes, hypertension, heart disease, and chronic obstructive pulmonary disease (COPD) have a huge impact on the healthcare system around the world. They do need continuous care, monitoring and prompt action to prevent complications and improve the patient outcome. Typically, people have a chronic condition managed through scheduled checkups and reactive treatment when symptoms arise. But AI is changing this—adapting to the proactive and personalized approach towards providing healthcare was successful. AI uses machine learning models and sophisticated data analysis to discern pattern in patient data, be able to predict the health deterioration and suggest specific intervention. AI can use electronic health records (EHRs), wearable device data, lab results and medicine history to predict complications and enlighten healthcare providers to keep away from emergencies and take benefit for excellent long-term patient health. Chronic disease management based on AI reduces hospital admissions, improves treatment, and improves patient care at large. Some key information given below:

- **Data-Driven Insights for Early Intervention**

  Early detection of health risks, as well as timely intervention, is possible on account of the ability for AI to process and analyze large volumes of patient data at a real time scale. Machine learning models can learn the fine pattern of the data which may represent early development of the disease progression or complications. For example, AI can calculate glucose levels, eating habits, and adherence to medication to foresee hypoglycemic or hyperglycemic episode in diabetic patients. Just like this, AI can monitor patients with heart problems for erratic heart rhythms and high blood pressure to alert the healthcare providers before a major health crisis occurs. Doctors have the opportunity to alter treatment plans, prescribe doses of medication, and recommend ways of living in order to avoid complications as a result of AI's predictive abilities. It also helps patients to keep stable health conditions and it decreases the likelihood of emergency hospital visits.

- **Customized Care Plans Based on Patient Profiles**

Since patient's unique health profile (that includes genetic factors, personal medical history and lifestyle behavior) can be analyzed, AI helps improve the creation of individualized treatment plans from it. In contrast to one-size fits all, AI based models can suggest most appropriate medication and the therapy decision for a person based on person's genetic makeup and its previous treatment response. Pharmacogenomics is an example of when AI can predict which drugs a patient is most likely to respond to with few side effects. The treatment plans are also dynamically adjusted in relation to lifestyle factors, such as diet, exercise pattern, and stress levels, which it also takes into account. Virtual health assistants and mobile app progress this process by collecting real time patient data and recommendation. The tailor approach improves the effectiveness of treatment particular essay, lowers the side effects and fosters patients to adhere to their care plans better.

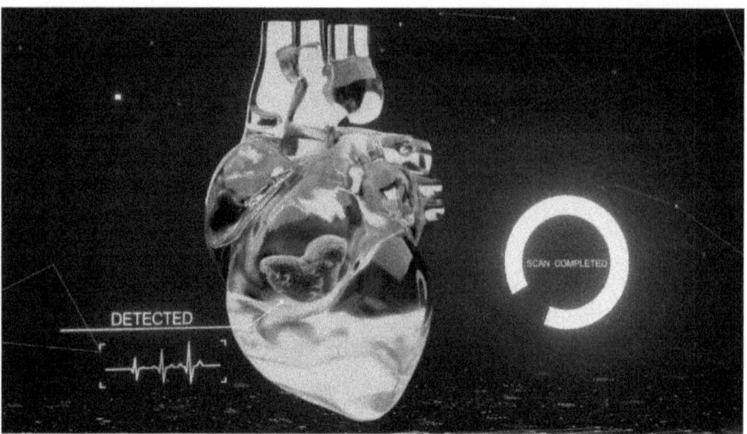

Figure 2.6: Cardiac Scan Results.

Source: (Zeleznik et al., 2021)

- **Continuous Monitoring and Real-Time Adjustments**

  Remote monitoring systems that use AI can track the health of the patient in continuous time even in conditions away from the clinical setting. Heart rate, blood pressure, oxygen saturation, sleep patterns, among other real time data, can be collected by the wearable devices smartwatches and health trackers. This data is analyzed by AI algorithms and found for absorbed anomalies and pre warning signals of health deterioration. For example, when an AI system monitors a heart disease patient, it can detect an irregular heartbeat and notify the patient as well as the healthcare provider immediately. On the other hand, continuous monitoring also assists AI by making real time adjustments of patient treatment plan according to the patient's changing health status. AI can also suggest medication adjustments or dietary changes in a patient with hypertension if blood pressure suddenly rises. This real time response reduces the chance of health emergency and enhances the patient outcome.

- **Empowering Patients Through AI-Driven Support**

  The use of AI empowers patients by providing them with tools and tools that facilitate action to their chronic conditions. Mobile apps that leverage AI help you with medication reminders, health suggestions and have a capability of providing you with personalized feedback that is based on the information of your running health. NLP means that AI can take patient reported symptoms and return tailored responses and recommendations. A diabetic patient who avails a virtual assistant that notifies them having high glucose levels can easily get certain suggestions like how to adjust the diet or change the medication. In addition, AI enables patients to be motivated in that it can track progress and provide positive reinforcement when health goals are achieved. AI endows patient with knowledge and real time support, which helps increase patient adherence to regulation plan, promotes healthy behaviors and improves general disease management.

2.3 Telemedicine and Remote Healthcare Solutions

Remote health care solutions and telemedicine have become an integral part of the modern-day healthcare as a way to provide medical services and to access them. The advancements in artificial intelligence (AI) have been rapid, the internet connectivity has improved and smart devices are becoming more and more prevalent, which has made it possible for patients to access medical care without the need for an in person visit. Telemedicine helps patients get in touch with the healthcare professionals through video calls, messaging apps and virtual health portal, and this is because the level of access to healthcare is now fast and efficient. This transition to some remote care has been most helpful for people in rural or underserved areas where specialist care can be a challenge.

Telemedicine, and remote healthcare in general, could not be effective without reliance on AI to increase diagnostic accuracy and enable assessment of real time symptoms, among other things, to support the care provided by healthcare providers. Virtual consultation platforms with AI underlying powered are capable of processing patient symptoms, medical history and real time health data and make accurate diagnosis and suggestion of appropriate treatment. Additionally, remote patient monitoring systems made with AI can be worn as wearable devices or smart sensors and track patient health metrics 24/7 and notify health care providers if we got a blood pressure spike before it becomes a problem. It means early intervention and reduces the chance of being taken to emergency hospital.

Secondly, remote healthcare solutions also help in improving healthcare efficiency through automation of patient triage and streamlining administrative tasks. AI algorithms can determine the urgency of a patient case, prioritize the patient cases that are critical, and direct patients to the level of care that is most appropriate for their condition. Furthermore, AI systems have the ability to complete language processing (NLP) and language translation capabilities, achieving overcame of the communication barriers for patients to receive the accurate information and the guidance, in the case of language or location. AI powered telemedicine and remote health solutions offer the benefit of delivering medical services by making rural people access care, enhance diagnostic accuracy, enabling continuous health monitoring, leading to improvements in patient outcomes all over the globe.

2.3.1 AI-Powered Virtual Health Assistants

Virtual health assistants powered by AI are becoming an integral part of telemedicine and remote healthcare solutions that ensure high quality, availability and efficiency of patient care. Intelligent digital tools these are such which make use of superior artificial intelligence (AI) including natural language processing (NLP), machine learning, and real-time data analysis to give personalized assistance to patients with their healthcare. Included in mobile apps, web portals and smart home devices, virtual health assistants support patients at anytime, anywhere, helping them with medical guidance provided by a health assistant. It can answer all medical questions, and give information on medication and treatments, and even help patients through symptom assessment. Virtual health assistant are able to monitor patient health in real time, track vital signs and alert both the patient and the healthcare provider if anything unusual or suspicious make an appearance. Real time monitoring at this level makes sure that patients receive timely care or intervention which is very important in the case of chronic conditions and the prevention of complications (Stoumpos et al., 2023).

Figure 2.7: Virtual Doctor's Visit.

Source: (Mammen et al., 2018)

AI powered virtual health assistants also contribute to the efficiency with which healthcare is delivered by automating a number of routine and time-consuming tasks that would otherwise be performed by the health care providers. For instance, they can receive the initial description of symptoms, medical history and lifestyle information before a virtual consultation starts. The use of machine learning model analyzes this data to find the severity of symptoms and suggests probable diagnosis or treatments. It allows healthcare professionals to concentrate on critical care, reduce wait time and enhance total care delivery. At an additional level, AI virtual assistants can offer continuous patient engagement in the form of reminders to the patient to take medication, coordinate follow up appointments, and helping them to adhere to the recommended treatment plan. They also provide evidence-based recommendation to patients that are tailored to the patient's unique health profile in order to help patients make informed decisions about their health.

Additionally, virtual health assistants are developed with the capacity to offer multilingual support and eliminate communication obstacles which enable patients of different linguistic and cultural backgrounds to receive appropriate and clear medical advice. Because these assistants can understand and response patient queries in many languages through natural language processing and because the sentiment analysis is driven by AI these assistants can more empathetic and more effective communication when they detect signs of patient distress and confusion. The virtual health assistants also aid mental health support by providing stress management techniques, guided breathing exercises and mental health resources. Virtual health assistants serve as a life line of sorts in remote healthcare settings where supply of specialized care can be limited, bringing consistent, suitable medical guidance into reach of patients.

The remote healthcare and telemedicine is significantly impacted by use of AI powered virtual health assistants in that it automates administrative tasks, improves diagnostic accuracy, and improves patient engagement. It brings healthcare providers to extend their reach, manages larger patients more effectively, and offers high quality care by utilizing their resources in a better manner. This transition to the AI-supported healthcare ensures that patients receive personalized, timely and accurate medical support which enhances patient satisfaction as well as the overall health outcomes.

2.3.2 Wearable Technology for Health Tracking

Telemedicine and remote healthcare solutions have become inseparable from wearable technology for health tracking, which allows continuous monitoring of patient's health outside the traditional clinical setting. Real time health data includes heart rate, blood pressure, oxygen saturation, sleep patterns, activity levels, and even stress levels are collected by the devices such as smart watch, fitness trackers and biosensors. In these devices, they embed in advanced AI algorithms to analyze the collected data to identify the trends, detect abnormalities and gives actionable insights. For example, an AI intelligent smartwatch can check if there is an irregular heart rhythm and inform the user to seek the medical help immediately before any serious cardiac event. Wearable technology allows patients and healthcare providers to monitor chronic cases, manage acute symptoms and actively monitor for early warning signs of potential health issues all the time.

Figure 2.8: Wearable Health Technology.

Source: (Dinh-Le et al., 2019)

Wearable health data is of tremendous value but until now this has been unexploited due to the complexity of the patterns that can be derived from it and the need to quickly turn this data into action. Wearables are used to derive machine learning models, which predict health events such as heart attack, stroke, diabetic episode, which are measured via machines. For instance, continuous glucose monitors (CGMs) fitted with AI can monitor blood sugar levels and alert diabetic patients when they can increase or decrease, and take immediate actions accordingly. Besides this it also assists to make fitness and wellness plans better by studying how energetic we are, how effectively we go to sleep and how effectively our body recovers from a charge every day to endorse particular exercise plans and other changes in lifestyle. Additionally, wearable technology connected to telemedicine platforms facilitates the real-time access of patient data during virtual consultation by the healthcare providers, which imparts enhanced knowledge to the healthcare providers and puts them in a position to make more informed choices and design tailor made care strategies.

Wearing technology also aids remote patient monitoring which reduces the requirement of numerous in person visits and hospitalizations. Wearable AI could be used to automatically transmit health data to respective healthcare providers for monitoring the same in real time and making the necessary changes in a treatment plan. As such, people recovering from surgery can have smart patches that track wound healing, as well as detect infection signs and notify medical staff in the case of complications. Just as AI wearables can detect respiratory health deterioration - early warning signs of patients with chronic obstructive pulmonary disease (COPD) - AI wearables can monitor respiratory health in patients with the disease. That improves patient outcomes by allowing for timely medical intervention and lower threat to emergency situations.

Wearable technology is also being deployed to monitor your mental well-being and your emotional health other than physical health tracking. With AI wearables, these can measure the heart rate variability and the skin conductance to assess the patient's stress level and give them feedback on breathing techniques to help patients deal with anxiety. Sleep tracking functionality that analyses sleep cycles and suggests improvements for better sleep quality, will affect overall health and cognitive function. Combining wearable technology and AI along with the telemedicine, it provides an overall system for healthcare where personal data can be together with the predictive understanding, and then recommendations to the patients.

Wearable technology for health tracking empowers the patients to be an active part of their health management and lets the healthcare providers have access to the patient data continuously. Inevitably, this shift towards the use of AI in health monitoring decreases patient complications and hospitalizations, leading to overall savings in health care cost, particularly.

2.3.3 AI in Remote Diagnosis and Consultation

Remote diagnosis is impossible without AI, as it allows doctors to check patients accurately, quickly, and safely, without their having to visit the doctor personally. Usually, traditional diagnostic processes need numerous visits to the clinic, medical tests and a long time to evaluate and may delay treatment and therefore be expensive on the healthcare. Nevertheless, artificial intelligence (AI) has made integration of the same into the telemedicine platforms easy; where the healthcare providers can remotely assess patient symptoms, assess medical data and provide treatment options in real time. These AI-powered diagnostic systems use machine learning algorithms and natural language processing (NLP) to collect and interpret patient-reported problems, medical history and real time health from patient self-reported symptoms, electronic health record data (EHRs), and health worn on the body by the patient. This lets AI recognize patterns, spot risks of becoming sick and supply accurate diagnostic information that may take a bit more to find if you're a human healthcare practitioner. To illustrate, an AI system can monitor a patient's heart rate variability, sleep patterns and reported fatigue levels to detect preclinical markers of cardiac issues or sleep problems that might prevent them from being given a chance(Lee, 2021).

The remote consultation process is improved by AI which provides clinical decision support to the healthcare professional during virtual visits. The AI driven platforms will assist doctors in following up the diagnostic process by proposing conditions possible to the symptoms of the patient, highlighting inconsistencies in the medical history of the patient, and suggesting what tests and treatment are appropriate. For instance, if a patient says they are suffering from chest pain and short of breath, an AI system could compare these symptoms with real-time data from their wearable ECG device of the current heart attack or less serious thing like fear. It also automates patient triage by determining the level of severity of a given patient's condition and directing them to the appropriate level of care, either a virtual consultation, in person visit, or emergency care. For this, not only does it improve diagnostic accuracy but also helps in efficient use of the resources of the health facilities.

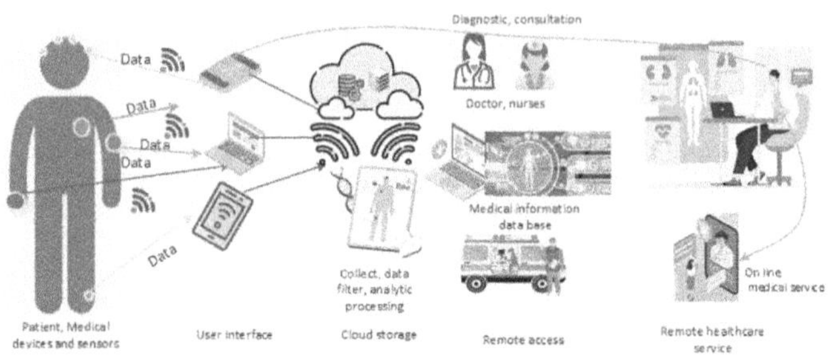

Figure 2.9: The main components Remote consultation.

Source: (Tsvetanov, 2024)

One of the best examples of using AI in remote diagnostic systems is for managing chronic diseases and complex health conditions. The monitoring of patient data over time can catch the slightest change that may indicate the path of a disease and the arrival of complications. By way of example: AI can study fluctuations of a diabetic patient's sugar levels and advise on adjustments to their medication or diet plans to avoid severe organs. Similarly, AI models can follow the pattern of breathing in COPD patients and warn the healthcare providers to sign of respiratory distress early. AI algorithms can also be used in the mental health care to uncover the signs of depression or anxiety during the virtual consultations by analyzing the speech, facial expressions and behavioral data.

One crucial strength of AI in aiding remote diagnosis is its ability to do multimodal data analysis — that is, analyzing medical images, lab tests results, genetic data, and patients' self-reported symptoms to provide an overview or a 'health profile' of a person. AI is able to do this, and this allows the AI to know complex correlations that may be missed in traditional diagnostic processes, to lead to an accurate diagnosis and successful treatment. Furthermore, AI assisted remote diagnosis provides for helps of personalized medicine and tailors treatment plans based on the features of a person. AI points out that by analyzing a patient's genetic makeup, lifestyle factors, and the earlier therapy, highly targeted therapies could get him or her to a successful outcome. The advantage of this personalized approach is that patients are given the most effective care with the least chance of an adverse reaction.

AI also enables continual learning and improvement in remote diagnosis. The AI systems become better at predicting and also at diagnosis the more cases they work and the more they receive feedback from healthcare providers. This is due to the fact that it guarantees the capacity of AI systems to learn themselves and therefore stay current with the latest medical research, treatment protocols, and patient outcomes so there can be continual improvements in delivering healthcare. In addition, remote diagnosis systems using AI are developed to overcome language and cultural barriers by incorporating natural language processing and translation ability to offer medical advice to a larger number of the patient populations worldwide.

The contribution of AI in remote diagnosis and consultation lies in boosting the accuracy of diagnostic, streamlining consultation process, and providing personalized care in order to revolutionize the healthcare landscape. Faster access to medical advice, earlier detection of health problems, more effective treatment and better-quality care — the patients benefit from and the healthcare providers can deliver high quality care faster. With the advancement towards AI-powered remote healthcare, apart from maintaining good patient outcomes, it also helps in reducing healthcare costs, eliminated preventable hospital visits, optimized the resource allocation and the pro-active disease management.

2.4 Healthcare Supply Chain and Inventory Management

Supply chain management (SCM) embraces all activities involved in the design, planning, execution, and management of an organization's supply chain. For a manufacturer, for example, SCM covers the processes of turning raw materials into finished products, as well as suppliers, logistics providers, distributors, and customers. A professional services firm's SCM, on the other hand, focuses on sourcing and managing the resources necessary to perform its consulting work. The most efficient supply chain management efforts try to integrate not only the discrete processes along the individual chain, but also the communication with supply chain partners that encourages cooperation up and down the value chain to minimize costs, maximize efficiency, improve quality, and ensure customer satisfaction.

What Is Healthcare Supply Chain Management?

Healthcare supply chain management emphasizes the strategic coordination of procurement and distribution of the goods required to sustain both business operations and the care of patients. Unlike industries that can zero in on a narrower range of related products, a healthcare supply chain needs to account for particularly stringent regulatory and compliance standards and could include everything from pens, laptops, and printing paper to perishable drugs and blood, X-ray machines, and surgical implants. Each healthcare organization must manage a broad network of vendors, processes, systems, and data and ensure that these items are manufactured, delivered, stored, and used so as to advance premium (and cost-effective) care to patients.

Key Takeaways

- Healthcare organizations need the right products at the right price at the right time to guarantee efficient patient care.

- Because healthcare organizations require such a wide variety of products, their supply chains tend to be vast and fragmented.

- The complexity of healthcare supply chains, along with supply chain disruptions, makes management especially challenging.

- Boosting data accuracy, automating processes, applying lean principles, and standardizing processes are just a few ways healthcare organizations can improve their supply chain management.

Healthcare Supply Chain Management Explained

Healthcare organizations engage an extensive, complex network of suppliers to keep operations running smoothly and deliver quality healthcare services to their patients. This process, known as healthcare supply chain management, aims to minimize costs and treatment delays while maintaining the highest standards of patient care and safety.

This is often easier said than done. A healthcare organization's supply chain is generally more complex and fragmented than those in other sectors, and to make matters more complicated, its management must take into account several high-stakes, industry-specific challenges, where lives are often on the line. However, the principles of good supply chain management remain largely the same from industry to industry: Even in healthcare, supply chain management encompasses not only oversight of vendors and products but also systems, processes, and data. This includes designing, planning, and managing the supply network, as well as and implementing mechanisms to predict, plan for, and react to possible disruptions or risks.

Healthcare Supply Chain Management Explained

Healthcare organizations engage an extensive, complex network of suppliers to keep operations running smoothly and deliver quality healthcare services to their patients. This process, known as healthcare supply chain management, aims to minimize costs and treatment delays while maintaining the highest standards of patient care and safety.

This is often easier said than done. A healthcare organization's supply chain is generally more complex and fragmented than those in other sectors, and to make matters more complicated, its management must take into account several high-stakes, industry-specific challenges, where lives are often on the line. However, the principles of good supply chain management remain largely the same from industry to industry: Even in healthcare, supply chain management encompasses not only oversight of vendors and products but also systems, processes, and data. This includes designing, planning, and managing the supply network, as well as and implementing mechanisms to predict, plan for, and react to possible disruptions or risks.

More specifically, healthcare supply chain management involves several key activities, including:

- Planning and forecasting demand for supplies, equipment, and medications.

- Collaborating with administrators and clinicians on product selection and standardization.

- Procuring and purchasing supplies from manufacturers and distributors.

- Managing and storing inventory.

- Distributing supplies to various facilities, departments, and locations.

- Tracking and managing product usage.

- Returning, repairing, or disposing of unused or expired items.

- Conducting quality control and testing of received products.

- Analyzing supply chain data to identify potential areas for cost savings and other optimization opportunities.

- Ensuring regulatory compliance as required, related to certain products.

- Managing relationships with suppliers and group purchasing organizations (GPOs), which aggregate the purchasing volume of multiple healthcare organizations to negotiate better prices, terms, and conditions with suppliers.

Healthcare Supply Chain Management Challenges

When healthcare organizations fine-tune their supply chains, they can achieve significant benefits: lower costs, increased revenues, and-most importantly-better care. But there is no shortage of challenges to overcome. According to Premier's 2024 Supply Chain Resiliency survey, nearly 80% of healthcare providers and 84% of suppliers expect supply chain challenges to worsen or remain the same over the next year.

These issues burden healthcare supply chain teams, providers, and patients. The same survey reveals that more than 67% of providers spend 10 or more hours per week mitigating supply chain challenges and shortages, and nearly 40% had to cancel or reschedule cases at least quarterly in 2023 due to product shortages-potentially putting patients at risk.

These situations, however, are not the only issues healthcare organizations face in managing their supply chains. Other all-too-common healthcare supply chain challenges include:

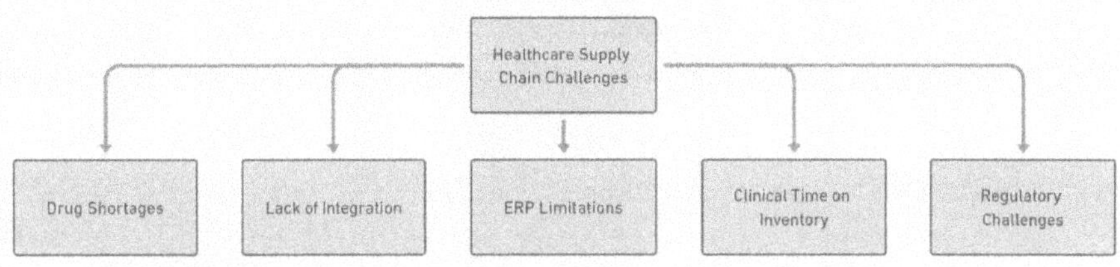

Figure 2.10: Challenges of health care supply chain

Source: (Self-generated)

- **Drug shortages:** Medication shortages have always been an issue in healthcare, but they've become a persistent and more frequent challenge due to ongoing supply chain disruptions, including global dependencies and demand surges related to pandemics or catastrophes. The U.S. Food and Drug Administration (FDA) notes that while manufacturing quality issues are the major cause of drug shortages, there are other culprits: production holdups, delays in receiving raw materials and components, and barriers to onboarding new suppliers during disruptions. Healthcare organizations that haven't taken proactive steps to mitigate these risks or diversify their supply chain often must resort to buying more expensive alternatives or maintaining costly backup medication inventory.

- **Lack of system integration:** The healthcare industry has seen significant merger and acquisition activity, which often exacerbates existing problems with supply chain system integration. When supply chain data exists in silos across an organization, meaningful analysis becomes difficult and real-time reporting next to impossible. Similarly, without centralized purchasing systems and channels, organizations will find it difficult to maintain companywide visibility into spending and supply chain management. The solution is system integration through linking disparate software applications or by investing in integrated enterprise platforms like cloud ERP. Benefits include greater efficiency, cost savings, business growth support, opportunities to automate processes, and, of course, the ability to make data-driven decisions.

- **ERP system limitations:** Healthcare's inherent complexity creates specific challenges for enterprise resource planning (ERP) implementation, particularly regarding regulatory compliance, data security, privacy rules, and integration with legacy systems and electronic health records software. While ERP software serves as the backbone of supply chain functions, healthcare organizations must ensure that their chosen systems fully support these specialized needs. Inadequate integration can force supply chain teams to manually transfer data between systems, which, in turn, drains resources, introduces errors, and becomes a drag on any attempt at supply chain automation, let alone optimization.

- **Clinical time allocated to inventory management:** An inefficient supply chain management function burdens both supply chains and healthcare providers. Poor inventory management, for example, often forces clinical staff to spend valuable time manually tracking and documenting products as they are received, stored, or used. That means that they're not devoting those hours to their primary role: patient care. The same is true for administrative staff, who may find themselves diverted from improving patient experience to handling supply chain issues.

- **Regulatory and compliance challenges:** Healthcare organizations must comply with a number of regulations that can affect their supply chains. The FDA oversees the safety and quality of medical products, which can impact sourcing, storage, and handling, Meanwhile, the data privacy and security regulations mandated by the Health Insurance Portability and Accountability Act (HIPAA) add complexity to supply chain management by requiring healthcare organizations to have comprehensive policies and procedures in place to control vendor representatives' access to physical facilities, as well as protections to prevent the unauthorized use and disclosure of patients' electronic protected health information.

2.4.1 AI for Drug Supply Chain Optimization

The drug supply chain management using Artificial Intelligence (AI) improves the accuracy, efficiency and responsiveness in pharmaceutical logistics. Pharmaceutical supply chain is complex with several stages in the drug manufacturing, warehousing, distribution and delivery to healthcare facilities and pharmacies. Fluctuating demand, supply chain disruption, inventory short, regulatory compliance and so on are the challenges for each stage. Traditional supply chain management methods tend to make extensive use of manual processes and static forecasting models that could easily make errors or errors. The AI provides an effective solution to streamline the drug supply chain operations by making the prediction, real time monitoring, automated decision making, and resource allocation optimal(Shashi Manish, 2022).

The AI driven systems can take large amount of historical as well as real time data patterns and predict the future of demand and automate the supply chain to adjust accordingly to the changes in market conditions and needs of patients. AI improves supply chain visibility while reducing lead time therefore helping a pharmaceutical company to remain more efficient and improve patient care outcomes. In addition, AI enables pharmaceutical companies to deal with the complexity of global supply chains, management of the supplier, and timely delivery of critical medications in times of great challenge, including natural disasters, political instability or global health emergencies.

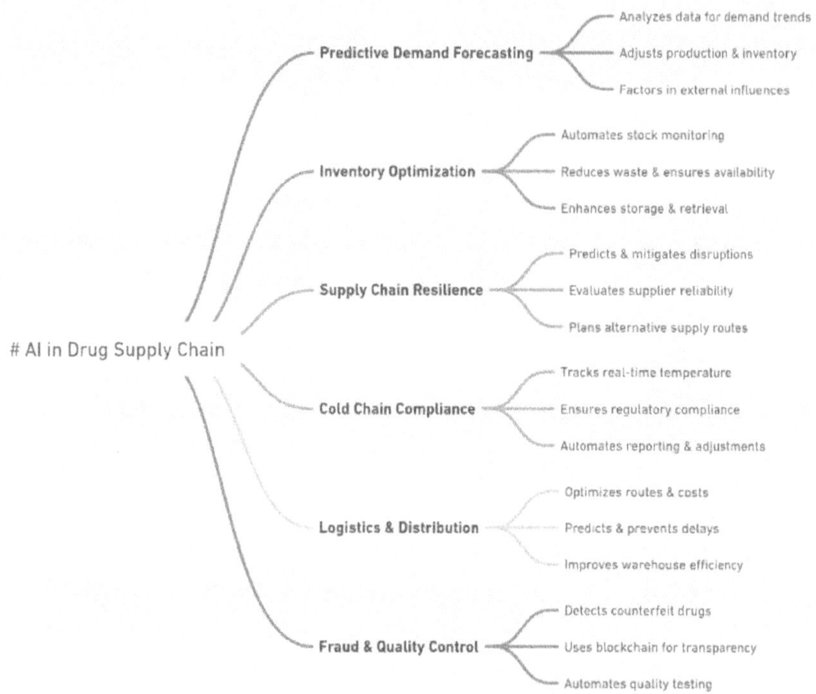

Figure 2.11: A Mind Map of AI's Role in Optimizing the Drug Supply Chain.

Source: (Self-generated)

1. **Predictive Demand Forecasting**

Predictive models based on historical data, market trends, patient population, and even seasonality are used to predict demands of drugs correctly using AI. With Machine learning algorithms, manufacturers and distributors can pre react to changes in the prescription patterns, diseases outbreaks and public health trends so that they can change production and inventory levels accordingly. For instance, during the flu seasons or pandemics, AI can predict around the demand for vaccines and antiviral drugs, in order to ensure there are no stock outs or a surplus of vaccines.

AI in pharmaceutical companies and healthcare services helps them to inform both the lack of stock and stockpiling of demand fluctuations. External factors such as political turmoil, trade environments, and the environment can also be accounted for by AI models in terms of drug supply and distribution. It provides a proactive approach to ensure availability of essential medication in all practical situations without waste and storage cost.

AI-based demand forecasting also assists the pharmaceutical companies make long term market changes. Take, for instance, supply chain management by which AI would analyze the population health data and epidemiological trends to predict its future healthcare needs and adjust production capacity to meet them. By allowing the pharmaceutical companies to be more agile and responsive to changes in global health issues, critical medications are available when and where they are needed.

2. **Inventory Management and Optimization**

Inventory management systems based on AI are monitoring the stock of drugs in real time and automatically replenish the drug stock. Inventory that is being used stale is utilized in the cigarette industry and machine learning algorithms take into consideration the inventory turnover rates, expiration dates and the storage conditions to maximize stock levels and minimize waste. Take as an example: AI can track how long a drug lasts on a shelf and schedule in the implicit use of those nearing expiration to avoid losses.

Demand driven inventory management is also supported by AI in adjusting stock levels according to the patient demand patterns and hospital usage rates. This dynamic approach results in high demand medications being readily available; low demand medications will be kept at the lowest possible holding costs. Automated inventory management systems also identify stockouts or surplus for automatic ordering or redistribution of drugs between healthcare facilities whenever they occur.

In addition, AI can work with hospital and pharmacy management systems to offer a global inventory snapshot of inventory levels throughout the entire supply chain. Pharmaceutical companies and healthcare providers can use this to coordinate inventory management efforts and avoid drug shortages at important points in the flow of products down the supply chain. Along these lines, AI based inventory management systems can further optimize warehousing operations by suggesting more efficient ways to store the inventory as well as reduced amount of time to retrieve inventory from the storage.

3. **Supply Chain Disruption Management**

Supply chain resilience is enhanced through AI's ability to anticipate and mitigate potential disruptions. Changes in regulatory environment, geopolitical conflicts, natural disasters, transportation failures, are the types of things that can have an effect on drug availability. Real time data coming from weather reports, shipping logs, geopolitical news, among others, are used to predict disruptions and recommend alternative routes or suppliers by AI models.

For example, if a manufacturing plant is in danger of closing because of a natural disaster, AI can quickly analyze the potential for the relocation of manufacturing sites and adjusting the supply routes. Supply chain managers can run different disruption scenarios through AI-driven simulation models in order to develop contingency plans in case of a crisis to maintain a drug supply.

Similarly, AI based systems help in supplier management by assessing the reliability and performance of the suppliers. AI is able to recommend the best suppliers for particular drugs and negotiate better terms with suppliers based on analyzing supplier delivery times, quality control data and contract terms. It helps pharmaceutical companies to reduce the supply chain risks and enhance the overall supply chain performance.

4. **Cold Chain Monitoring and Compliance**

Vaccines and biologics are many pharmaceuticals, that need specific storage and transportation conditions such as strict temperature control. Internet of Things (IoT) sensors placed in cold chain monitor AI based systems that constantly record and reports temperature, humidity and transportation conditions in real time. This data is then analyzed by machine learning algorithm to detect deviations and then the machine learning algorithm automatically changes the storage or shipping parameters to keep the conditions optimal.

It allows real corrective actions to be taken when storage or transportation conditions are compromised. Proper functionality to all regulatory requirements for storage and transport of drugs is vital for drug efficacy and safety of patients. Monitoring of drugs with AI is so driven so that the drugs meet the quality standards and remain efficacious along the supply chain.

Also, AI has made pharmaceutical companies comply with international regulatory standards regarding cold chain management. The obvious advantage of using AI is that it not only automates the generating of the automated compliance reports and details of storage and transportation conditions but also helps maintain detailed records of these storage and transportation conditions with which one can easily go through in a regulatory auditors audit and reduce the risk of being fined for noncompliance penalties for failure to maintain required records of storage and transportation conditions. Cold chain system could also be predicted by AI based systems on the effect of environment factors on cold chain performance, and suggested improvement to boost cold chain operation reliability in general.

5. **Logistics and Distribution Optimization**

AI suits drug transportation and delivery by exploiting other traffic patterns, transportation costs, and delivery timeline. Route optimization algorithms used are guidance algorithms to find the fastest and the most fuel-efficient delivery paths in lieu of street condition, weather and price of fuel. Just as AI can be used for consolidating deliveries to incur lower costs and MA minimize environmental impact in drug distribution, these systems can also help chase the order to minimize the work related to deliveries and hence the overall cost.

Deliveries delays of any nature can be predicted by Machine learning model and delivery delays can be recommended to corrective actions such as rerouting the shipments or deploying alternative carriers. AI based logistics platforms connect with the inventory management and demand forecasting systems and the end play out comes as a controlled ready to use supply chain network that fetches drugs to healthcare providers and the patients at right time.

These real time visibility of drug shipments are also enabled by AI driven logistics systems, helping the pharmaceutical companies and healthcare providers to achieve better visibility of the performance of the supply chain. AI systems can automatically initiate recovery procedures and alter the schedule of distribution so as to prevent the maximum possible disruption, if a shipment is delayed or damaged. AI based logistics can also suggest the strategic warehouse location as well as distribution hubs that will reduce the transportation cost and also improve delivery time.

6. Fraud Detection and Quality Control

AI helps to secure drug supply chain through detection and identification of counterfeit drugs and supply chain fraud. The algorithms that can detect whether something is counterfeit using machine learning will analyze shipment records, supplier information, and product codes. The effect of AI based blockchain solutions is further enhanced with transparency as they provide an immutable chain of drug manufacturing, distribution and delivery.

Then artificial intelligence systems are used to analyze drug samples and production data in order to see if any composition or potency can vary. Quality testing is automated to minimize human error and to guarantee only high-quality drugs in the supply chain. Pharmaceutical companies are able to assert their regulatory compliance and protect patient's safety through this.

AI based fraud detection systems can also let you know about suspicious trend in supplier contract, payment records and drug pricing data. AI aids pharmaceutical companies in preventing financial losses and maintaining integrity of the drug supply chain with the help of detecting anomalies and flagging potential fraud cases. AI systems can also scan the social media and online marketplaces looking for reports of counterfeit drugs, giving quick response in case of an emerging threat.

2.4.2 AI in Predicting Medical Equipment Demand

Machine learning algorithms, real time data analysis, computing predictive model and using it on medical equipment demand prediction is what AI has revolutionized. Depending on fluctuating patient needs, seasonal variations, and unforeseen emergencies such as pandemic, traditional forecasting methods do not perform well in accuracy. Using AI powered systems that turn through many datasets, such as hospital admission rates, disease prevalence, historical usage patterns, real time epidemiological trends, these systems predict spikes in demand for such critical medical equipment. AI integrates with electronic health records (EHR) and hospital management system to give healthcare administrators accurate demand projections that help them to proactively procure ventilators, diagnostic tools, surgical instruments and rehabilitation devices. It leads to this data-driven approach that helps hospitals avoid shortages, eliminate excess stockpiling and that critical equipment is on hand when it is required(Panesar, 2019).

In addition to forecasting, AI optimizes supply chain and logistics by discovering demand fluctuations both at a local and national level. With AI-driven insights, healthcare networks can use coordination plan of equipment's distribution, dynamic resource reallocation and routing of transportation to minimize delays. The use of AI inventory management also leads to lower operational costs because it keeps a good balance of the stock, over purchasing is avoided, and equipment depreciation is reduced. AI can propose the best time for the hospitals to purchase expensive medical devices and save costs, by analyzing the supplier trends and market conditions. With the continual progress of AI technology, AI will find its way into healthcare supply chains to make healthcare supply chains more resilient by guaranteeing that medical facilities will always be stocked with enough medical supplies to meet patient needs and cut down on waste and improve the efficiency of healthcare as a whole.

2.5 Automated Medical Documentation and Workflow Optimization

AI driven technologies also play an important role in the automated medical documentation and workflow optimization to enhance efficiency, accuracy as well as productivity that is being used in healthcare settings. By combining natural language processing (NLP) and machine learning, increased AI powers can transcribe physician patient interactions, extract most beneficial medical insights and create well-structured clinical notes. They free up healthcare professionals' time from administrative burdens to spend more time taking care of patients instead of busy work. On top of that, AI driven workflow automation cuts hospital operations such as appointment scheduling, billing and medical coding down to a bare minimum resulting in efficient and less error prone workflow.

Apart from providing the documentation, this AI-powered workflow optimization ensures that there is no gap between healthcare providers and patients as well as support staff. This allows intelligent systems to analyze patient history, predict care needs and can provide personalized treatment plan, to improve the ability to make clinical decision. They increase compliance with the treatment protocols for a reduced chance of missed diagnose or medication errors through automated alerts and reminders. Meanwhile, AI built tools sync with the electronic health records (EHRs) so that knowledge can be shared in real time across departments; so collaboration is more accessible and continuity of care is achieved. With AI developing in the next few years, there will be a great need for automating medical documentation and optimizing the workflow to reduce burnout of the healthcare professionals and perform better for the patients.

2.5.1 AI for Clinical Note Generation

Clinical notes are a critical part of the behavioral healthcare, capturing a complete history of the patient, the plan of care and the outcome of the patient. These notes are important as a crucial communication tool between healthcare providers team for the continuity of care and decision making. But manual note taking is a time consuming and laborious activity that providers lose time to focus on patient care. Documentation can add an administrative burden that can make the clinician even more burned out, less efficient or delay a clinician's ability to accurately update patients' records.

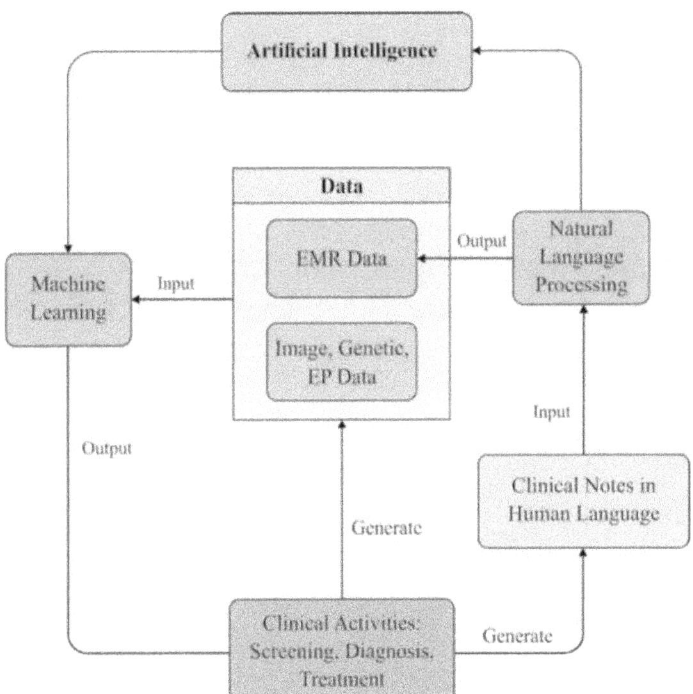

Figure 2.12: Roadmap of clinical data generation.

Source: (Colonna, 2021)

Behavioral healthcare providers are able to make use of an effortless, efficient solution to their problem with manual note taking with the introduction of AI powered clinical notes generation. And thanks to a combination of natural language processing (NLP), machine learning, they can automatically transcribe patient interactions, extract important bits and punch them in writing in real time. Not only it improves workflow but also it simplifies workflow, improves accuracy, reduces errors and conforms to regulatory standards. For that reason, providers can spend more time with patients and also get comprehensive and organized records. Here are some benefits of AI in clinical note generation:

- **Increased Efficiency:** One of the primary benefits of AI-powered clinical notes generation is increased efficiency. Manual note-taking can be a time-consuming and tedious process, especially for providers who have a high patient volume. It can also be challenging to document everything in real-time, leading to incomplete or inaccurate clinical notes. With Auto Notes, providers can automate the clinical notes generation process, saving them time and allowing them to focus on patient care. Auto Notes accurately captures patient progress during sessions and generates comprehensive, compliant clinical notes in real-time, ensuring that providers have all the information they need to provide quality care to their patients. By reducing the time and effort required for note-taking, providers can see more patients, which can lead to improved revenue and a higher level of patient satisfaction.

- **Accuracy and Completeness:** Manual note-taking can also lead to incomplete or inaccurate clinical notes, which can impact patient care and compliance. Providers may forget to document important details or make errors when handwriting notes, leading to incomplete or inaccurate documentation. With Auto Notes, providers can have confidence in the accuracy and completeness of their clinical notes. Auto Notes captures patient progress in real-time, ensuring that nothing is missed, and generates comprehensive, compliant clinical notes that include all necessary details. This not only improves patient care but can also reduce the risk of malpractice and improve compliance with regulatory requirements.

- **Compliance:** Compliance is a critical consideration in behavioral healthcare, with strict HIPAA regulations governing the handling and storage of patient information. Manual note-taking can make it challenging to maintain compliance, as providers must ensure that all documentation is accurate, secure, and confidential. Auto Notes offers a secure and compliant solution for clinical notes generation, ensuring that patient information is kept confidential and secure at all times. Auto Notes maintains strict compliance with HIPAA regulations, providing providers with peace of mind knowing that their clinical notes are secure and compliant.

- **Improved Patient Care:** Ultimately, the benefits of AI-powered clinical notes generation can lead to improved patient care. By streamlining the clinical notes generation process, providers can spend more time focusing on patient care, leading to better outcomes and higher patient satisfaction. The accuracy and completeness of clinical notes generated by Auto Notes can also lead to improved patient care, as providers have all the information, they need to make informed decisions about patient treatment. With improved patient care, providers can build stronger relationships with their patients, leading to improved outcomes and higher patient retention.

2.5.2 AI-Powered Voice Recognition for Physicians

The AI voice recognition is helping to boost physician efficiency and remove burdens from the administrative aspects of the healthcare industry. Advanced natural language processing (NLP) based on deep learning models, used to make voice recognition systems, help physicians dictate clinical notes, prescriptions and patient histories with very high accuracy. They integrate perfectly with electronic health records (EHRs) and complete documentation without manual data entry error. Not only does it save time but, most importantly, doctors can dedicate more time on patient care rather than paperwork. In addition, speech to text capabilities in real time allows for real time note taking during consultation, improving communication and ensuring the critical details are not lost(Carrasco Ramírez, 2024).

But both voice recognition by AI and its application in clinical decision making and patient Relation are also being used beyond documentation. There are some systems capable of distinguishing medical terms, flagging possible errors and even recommending possible treatments on spoken input. AI voice assistants can assist physicians with hands free retrieval of patient's records, set LIS reminders and retrieve relevant medical literature. In addition, by integrating voice recognition to telemedicine platforms, remote consultations are facilitated with perfect voice-based interaction between doctors and patients. This is due to the fact that as AI technology develops, voice recognition will begin to be used more and more in healthcare to create a more efficient and accurate patient centric healthcare system.

How Does Voice Recognition Work?

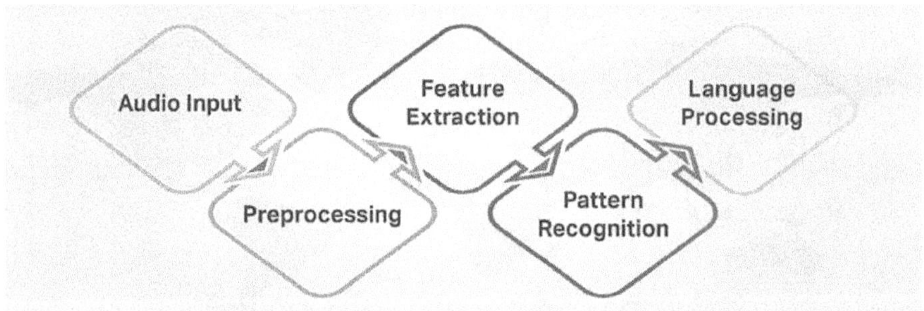

Figure 2.13: Audio Data Processing Stages.

Source: (Lhoest et al., 2021)

- **Audio Input:** The process begins with capturing the audio input using a microphone.

- **Preprocessing:** The audio signal is cleaned up by removing noise and normalizing the volume.

- **Feature Extraction:** The system analyzes the audio to extract key features such as pitch, tone, and frequency.

- **Pattern Recognition:** The extracted features are compared to known patterns of speech stored in a database.

- **Language Processing:** The recognized patterns are converted into text, and natural language processing (NLP) algorithms interpret the meaning.

2.5.3 AI in Standardizing Medical Records

A big role AI had in standardizing medical records is addressing inconsistencies, reducing errors and improving accuracy of the data in healthcare systems. Medical records in traditional ways are often fragmented, have inconsistent terms, and have human errors that impair effective patient care, research and data analysis. The variation in terminology and formatting in the compilation of medical records from several sources such as clinical notes, lab reports, imaging results and patient histories are common. The result is that these diverse sources of data can be fed into AI powered natural language processing (NLP) and machine learning (ML) models, which can extract, interpret and normalize, the data so that it is consistent and accurate. These systems are able to store medical terms, identify some patterns, and map the medical terms to a standard coding system like SNOMED CT (Systematized Nomenclature of Medicine – Clinical Terms) or ICD-10 (International Classification of Diseases). This automating this process with AI reduces having to hand write data, reduces human error, and improves the quality of medical records as a whole. Better communication among healthcare providers, increased reliability of patient records and ability to share the information between healthcare institutions are possible with enhanced data consistency and accuracy (Li et al., 2024).

Also, medical records standardization by an AI boosted clinical decision support, patient care and population health management. If medical records are consistently structured and formatted, healthcare providers have more ease analyzing patient histories, tracking disease progression as well as identifying treatment patterns. With standardized data, trained using AI powered predictive model which are able to identify patient risk factors, predict the potential complications, and recommend personalized treatment plan, with higher precision. One way is that AI can collate all of patient's medical history, lab results, and can analyze information about genetics of a patient to identify targeted therapies and alert about possible adverse drug interactions. Also, AI codes diagnoses and procedures automating the whole process and thus accuracy in billing and less work for the healthcare staff. It frees up health professionals to spend more time with patients than filling in forms. In addition, AI helps in data exchange and interoperability of healthcare systems by bringing records to the same standards. It improves coordination of care, facilitates more effective and efficient referrals to referring and receiving clinical disciplines within healthcare networks. Standardized medical records also facilitate a more rigorous and ecologically valid medical research that, in turn, helps additional development of new treatments and enhances public health strategies. With AI technology, it will get more standardized in medical records and this will aid in creating a better efficient, accurate and patient centered healthcare system.

Over all, Its cover how AI is being used to change how healthcare gets run and operate in terms of being more efficient, more accurate and better improving overall care for patients. The optimization of hospital administration includes using advanced patient scheduling and resource allocation to optimally schedule the maximum number of patients for making the best use of hospital staffing, equipment, facilities. Automation tools relieve administrative hurdles from executing same tasks like billing and record management; these same tasks are handled by the tools, leaving the healthcare professionals with more time for patient care. Predictive analytics increase patient care by predictive of disease, readmission risk and chronic disease management through personalized treatment plans. At the same time, AI is providing assistance to telemedicine through virtual health assistants and remote diagnostic tools that give real-time patient insights and better provide consultation. In healthcare supply chains, AI provides predictions about drug and equipment demand, thus decreasing some drug and equipment shortages and improving inventory effectiveness. Then, AI driven stories simplify the medical documents by eliminating the manual clinical document generation, doing real time voice transcription or standardizing medical record to make data consistent and interoperable. AI is making a big difference in healthcare delivery and patient outcomes with the help of the AI that is increasing the operational efficiency and clinical decision making.

Multiple Choice Questions (MCQs)

1. **What is one primary benefit of using AI for patient scheduling and resource allocation in hospital administration?**

    a. Reducing the number of patients

    b. Increasing the length of hospital stays

    c. Efficient management of staffing and equipment availability

    d. Limiting the number of available resources

2. **How does AI reduce administrative burden in healthcare operations?**

    a. By increasing manual data entry

    b. By automating tasks such as billing and appointment scheduling

    c. By replacing healthcare professionals

    d. By reducing patient care time

3. **Which of the following is a key advantage of using AI for early disease prediction and prevention?**

    a. Reduced need for medical staff

    b. Identification of disease patterns and patient risk factors

    c. Delayed diagnosis of diseases

    d. Elimination of follow-up care

4. **AI in hospital readmission risk analysis helps healthcare providers by:**

   a. Predicting which patients are likely to be readmitted

   b. Reducing the number of hospital beds

   c. Encouraging longer hospital stays

   d. Eliminating patient records

5. **How does AI support chronic disease management?**

   a. By identifying trends in patient data and suggesting personalized treatment plans

   b. By reducing access to patient records

   c. By increasing the workload of healthcare providers

   d. By limiting the use of predictive models

6. **What role do AI-powered virtual health assistants play in telemedicine?**

   a. They diagnose diseases without human input

   b. They provide real-time guidance and support to patients

   c. They replace physicians in consultations

   d. They monitor only physical activity

7. **How does wearable technology contribute to healthcare operations?**

   a. By recording patient data for research purposes only

   b. By providing continuous health tracking and early detection of health issues

   c. By limiting data collection to hospital visits

   d. By increasing the need for patient monitoring in hospitals

8. **AI for drug supply chain optimization primarily helps by:**

    a. Increasing drug production

    b. Predicting drug demand and reducing shortages

    c. Eliminating the need for medical inventory

    d. Reducing drug costs directly

9. **How does AI assist in clinical note generation?**

    a. By transcribing physician-patient interactions into structured notes

    b. By eliminating the need for documentation

    c. By requiring manual entry for all data

    d. By delaying the recording of patient data

10. **Why is AI-based standardization of medical records important?**

    a. It reduces the need for patient history

    b. It ensures consistency and improves data interoperability across healthcare systems

    c. It eliminates the need for medical coding

    d. It restricts access to patient information

Answers:

1	2	3	4	5	6	7	8	9	10
c	b	b	a	a	b	b	b	a	b

Challenges and Ethical Concerns in AI Healthcare

3.1 Data Privacy and Security in AI-Enabled Healthcare

Healthcare systems powered by AI are designing AI enabled systems which use large volumes of sensitive health data to analyze and improve health diagnosis, health treatment and healthcare management. This kind of data is PHI (personal health information) and includes medical histories, lab results, imaging data, genetic profiles, and treatment plan; all very sensitive information that needs to be protected very rigorously. Here's a need to always ensure privacy and security of data as AI systems become more and more easy and integrated their work into health care operations. Data breaches, unauthorized access, data misuse provide potential risks to trust due to AI models being dependent of big data sets for training and operating AI models. Healthcare organizations are under the threat of cyberattack without proper safeguards that can lead to data theft, data manipulation, as well as ransomware and can put patient safety and trust on the line (Pezoulas et al., 2020).

The solution for this problem is the strong encryption protocols, secure data storage solution and robust access controls employed by healthcare organizations. Encryption ensures that the patient data is protected in transmission and storage while access controls ensure that any patient data is available only to the authorized personnel. Securing AI based healthcare systems is become essential, and require at the very advanced techniques; such as differential privacy and federated learning. Differential privacy is the method of protecting patient anonymity by adding noise to datasets in such a way that individual details cannot be traced back to specific patients. One of the new key features of federated learning is training of AI models without transferring patient data to a central server, which minimizes the risk of exposure, especially encompassing the question of adequate privacy.

Data privacy and security in AI healthcare also means complying with the healthcare regulations. Based on such regulations such as the Health Insurance Portability and Accountability Act (HIPAA) in the United States and the General Data Protection Regulation (GDPR) in the European Union, capturing, storing and processing of the patient data is to be done strictly as per regulations. These regulations make it clear that AI systems must be designed in such a way that patient consent is obtained before data usage and data processing is transparent and secure. Regular audits and security assessments by the healthcare providers are also needed to identify vulnerabilities and also for compliance to the evolving data protection standards.

In addition, cyber security tools such as multi factor authentication, real time threat detection and their automated response tools become important for protection of AI based health care platforms from hacking and data breaches. The machine learning algorithms can also be used to detect unfair activity and potential threats to security and be faster at reacting when security leaks occur. Furthermore, as data are a fundamental part of the AI driven healthcare, integrity of the data is important as they can result in misdiagnosis and compromising patient care. Integration of robust privacy and security into AI enabled healthcare systems can increase patient trust, ensure confidentiality of data, and give better and more reliable medical care by the healthcare providers. Successfully gaining an AI driven healthcare system becomes a technological and an ethical responsibility to protect patients' rights as well as better healthcare outcome.

3.1.1 AI in Securing Electronic Health Records (EHRs)

Electronic Health Records (EHRs) are the backbone of modern healthcare – a place to store and manage patient's information. EHRs contain a highly sensitive data including names, medical histories, diagnostic reports, medication records and personal identifiers. Faced with both increasing digitization of healthcare and accompanying challenges to security, the forces of digitization have enabled the provision of better patient care and improved operational efficiency to healthcare organizations, but have also posed challenges to security. More sophisticated security measure in the healthcare organizations has become necessary since data breaches, unauthorized access, and cyberattacks have risen sharply against the healthcare organizations. These traditional security approaches like static encryption and password based access no longer suffice in ensuring a safe environment for the ever increasing volume and complexities in the volume and complexity of that data(Pepper, 2023).

The security of EHR systems is being improved through the emergence of Artificial Intelligence (AI). In fact, AI driven solutions can help you analyze plenty of data put through, detect anomalies, can automate encryption and enforce access control in real time. The machine learning models make it possible to identify the suspicious behavior such as the unauthorized access attempts or odd data flow and respond to the threat in a more timely and accurate manner than the human operated systems. In addition to the advances in encryption, behavior base, authentication, and automated compliance monitoring, AI allows the healthcare provider to continue to meet operational needs by protecting sensitive patient data. In this chapter, the author explores what are the main ways by which AI is used to protect EHRs, i.e., threat detection, encryption, access control, data integrity, and regulatory compliance.

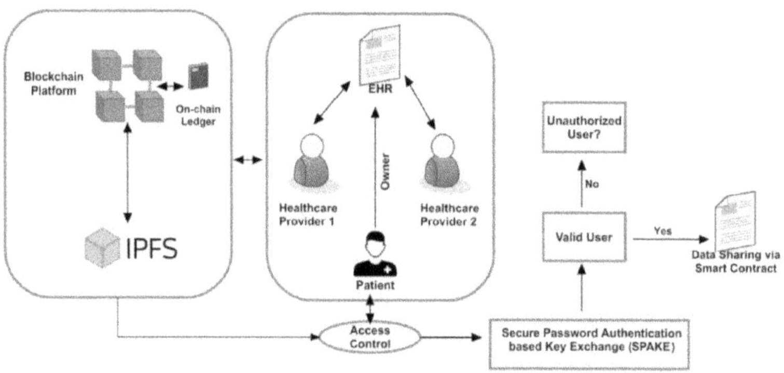

Figure 3.1: Blockchain-Based EHR Sharing System.

Source: (Sonkamble et al., 2023)

AI-Powered Threat Detection and Anomaly Recognition

EHR security has been quite enhanced by the boost of AI powered threat detection and anomaly recognition for real time monitoring and the reaction to the suspicious activities. The ability of machine learning models to analyses huge amounts of historical and real time data to identify such patterns, like large data exports, unusual login attempts or unauthorized access to sensitive data, is hence also very effective in identifying potential breaches. The applications of deep learning algorithms is that they can detect very minute changes in the user behavior that regular security systems cannot and healthcare organizations can act quickly to arms to these possible hours. Security Information and Event Management (SIEM) systems based on AI can correlate many security events to detect complex and coordinated attacks to increase the overall resilience of EHR infrastructure.

- **Advanced Encryption and Data Masking:** AI has also helped in encrypting and masking the data both in rest and in transit. Internet cryptography is generally built on static key management, thus making them vulnerable under sophisticated attacks. Encryption systems that are driven by AI can change encryption protocol dynamically on the basis of the sensitivity of data, the nature of the network environment.

- **Homomorphic Encryption:** AI models can run against patient data in an encrypted format, through homomorphic encryption, so it's not exposed to potential breaches as it's being used. It makes sure that even after analysis, the data is not given out unless it includes only sensitive information.

- **Dynamic Data Masking:** It also has the capability of dynamic data masking: it hides sensitive information not visible to unauthorized users but allows for the operational access to non-sensitive data. It guarantees that the confidentiality of the patient is not compromised while allowing access to the required medical information.

- **Automated Access Control and Identity Verification:** New security to EHR systems was provided by the AI driven access control and identity verification systems. Usually, traditional password-based systems are easy to hack and are also open to social engineering attacks. Biometric systems based on AI software used for identity verification (facial recognition, fingerprint scanning and voice authentication) to prevent unauthorized personnel accessed to EHRs.

- **Behavior-Based Multi-Factor Authentication:** Behavior-based multi factor authentication (MFA) is something that AI can implement as well, configuring access parameters based on a relevant contextual factor such as the location, device type and user behavior. The security is strengthened by shaping authentication needs to match the risk.

- **Insider Threat Detection:** Professionals in the AI discipline examine user behavior in real time, tracking when deviations from the typical behavior occur and use this to detect any insider threats. The AI system allows the user to restrict access if an authorized user tries to access sensitive data outside of their usual activity.

- **Ensuring Data Integrity and Tamper Detection:** Another important function of AI in EHR security is ensuring the integrity of the data. An immutable audit trail can be created by AI models to track and log every change made to EHRs. Blockchain based AI systems take further processing on the integrity of data through the uncompromising, cryptographically secured, decentralized, structures of transaction records.

- **Inconsistency and Tampering Detection:** AI can also check the consistency of records or spots of illegitimate change, as well as automatically reconstruct lost or damaged data by spotting certain patterns in history. It guarantees that after security incidents, patient records are accurate and trustworthy.

3.1.2 Cybersecurity Risks in AI Healthcare Systems

The integration of Artificial Intelligence (AI) in the healthcare system has brought forth a lot of developments in the patient care, diagnostics and operational efficiency. At the same time, it has introduced new vulnerabilities to sensitive patient data that are very serious. But, as the healthcare system is dependent heavily on AI driven healthcare systems that involves big volumes of personal and medical information, they constitute prime targets for the cyber criminals. Among the most usual threats are data breaches, unauthorized access, ransomware attacks, and other breaches. In particular, machine learning models, which will need access to large datasets for training and operations are highly at risk for model poisoning and data manipulation. Model poisoning attacks involve poisoning the training set with malicious data, which results in wrong predictions and leaking out wrong information to patients. In addition to that, adversarial attacks against AI systems are also threatening by manipulating input data—e.g. introducing the problem of adversarial attacks where a small, nearly imperceptible perturbation is added to the data input to misguide the AI model into incorrect diagnosis or treatment recommendations. These risks underline the need to make AIs safe from both external and internal vulnerabilities, and external threats.

Like in any other application of AI, AI healthcare systems also have their own challenges when it comes to system integrity as well as regulatory compliance. AI Model Drift or when AI model's performance begins to degrade gradually because of changes in input data patterns, can undermine diagnostic accuracy and contributes to wrong decision making. This allows cybercriminals to introduce misleading data to disrupt AI model performance. Another threat is insider threats in which authorized personnel abuse their access authorities. Yet growing AI models are still difficult to regulation compliance, since healthcare organizations have to abide by strict data protection laws such as HIPAA and GDPR. A failure to comply can bring about legal fines and damage to patient trust. To address these cybersecurity risk, it requires multi layered approach with encryption, behavior-based detection and adaptive access control. To secure AI healthcare systems and protect patient's data confidentiality & integrity, one needs to have Proactive monitoring and continuous AI Model updates and Real time response systems.

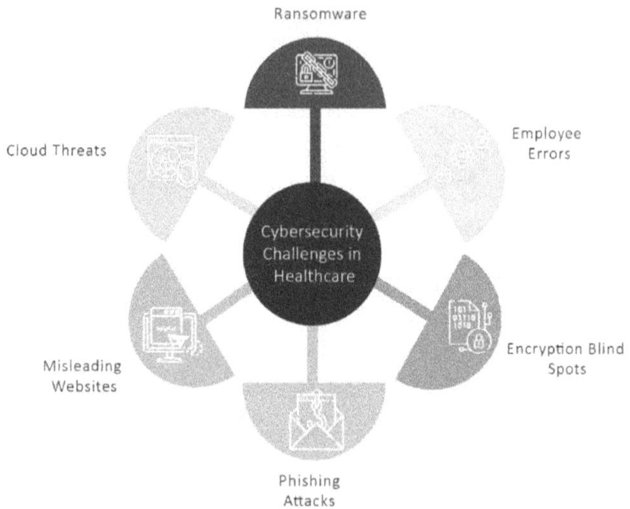

Figure 3.2: Main Cybersecurity in Healthcare Challenges.

Source: (Sendelj & Ognjanovic, 2022)

1. **Ransomware** – A ransomware attack gets records hacked using malicious software that will then encrypt healthcare data that can only be decrypted at a price. If they are not paid, these attacks can disrupt critical health care services, prolong patient care, and it can reveal sensitive patient information.

2. **Employee errors**- Human error is also a major vulnerability in healthcare cybersecurity: at all levels from Employees. Weak password, accidental sharing of sensitive data, or in handling of access credentials are mistakes that can expose sensitive data or compromise the system.

3. **Encryption Blind Spots** – Encryption is a must to protect patient data but blind spots are when certain flows of data or storage points are not encrypted. These gaps are exploitable for attackers to intercept and steal sensitive information.

4. **Phishing Attacks** – Phishing entails the fraudulent emails and messages that persuade the healthcare staff to compromise access to the login credentials or any of the other sensitive

information. Phishing attacks can be really successful and they can eventually result in unauthorized access to the system and the data breach respectively.

5. **Web of Misleading Websites** – Cybercriminals make fake websites similar to legitimate healthcare portals that lures users to enter sensitive info on it. They can be used to steal login credentials, financial details or patient records.

6. **Increased Healthcare Use of Cloud Technologies** – The more we use cloud technologies in and for healthcare, the more we will be exposed to their vulnerabilities that could lie in the cloud infrastructure. Data can be exposed and there is unauthorized access because cloud settings are not configured correctly, there is no encryption, and insecure APIs exist.

3.2 Bias and Fairness in AI Medical Applications

Artificial Intelligence (AI) is being utilized for medical applications to bring a significant change in it due to its efficiency in improving diagnostic accuracy, personalizing the treatment plans, and boosting operational efficiency. Yet, right now radiology uses AI systems, there are AI systems in radiology pathology and drug discovery, and there are AI systems. Although such developments have happened, they also allow biases that serve to make the healthcare in straight far from the people in charge of it to take over even further in the future. Seldom is training data (when available) balanced or complete resulting in a bias in AI medical application responsible for lacking for example the minorities, the older adults, women and underserved communities. A problem with an AI model is that it might not perform as well when it's applied to patients outside of the group that it was trained on. Assume that a dermatology AI model that has been trained mostly on lighter skin tones will be less accurate for identifying skin conditions in people with darker skin tones, misdiagnosing and failing to provide appropriate treatment. Likewise, AI models for cardiovascular risk assessment have been demonstrated to fail to account for risks for patients from some of the racial and ethnic groups that are underrepresented in the training data. This highlights that such improvements in the generalizability and accuracy of AI models should rely on diversity and inclusiveness in training datasets and thus require them to be seen to become generalizable and accurate over diverse populations(George et al., 2023).

AI medical applications are both a technical and an ethical and legal responsibility, and thus fairness in AI medical applications is not a technical challenge alone. AI systems should also deliver consistent, accurate results to different population. Tools of various kinds for algorithmic auditing and bias detection can assist the identification of performance disparities in the model and suggest means for remediation. For example, one such remedy to tail off the inherent biases in the AI system is through reweighting data samples, changing model architectures and applying of adversarial debiasing methods. However, explain ability and transparency in the decisions of AI systems are necessary for implementing trust in patients and healthcare providers. Given the importance of cancer diagnosis, drug prescription and surgical planning, clinicians need to know how AI models come to their conclusions. The lack of transparency can make one hesitate to adopt the AI based solutions, especially when it seems that the solution only favors some of the groups of patients. Healthcare organizations are required to make sure that the AI driven decisions are explainable and non-discriminatory, as per the frameworks such as the Health Insurance Portability and Accountability Act (HIPAA) in the US, and the General Data Protection Regulation (GDPR) in Europe. Therefore, AI development lifecycle has to incorporate fairness auditing tools as well as bias mitigation strategies applicable to data collection, model training, deployment and real-world performance evaluation.

Besides overcoming the technical and regulatory challenges, fairness in the use of AI in medical applications requires the cooperation between AI developers, healthcare professionals, ethicists, and policymakers. Including disparate ideas during the development stage makes it easier to detect a possible source of bias and make more reasonable solutions. In the case of creating transparency and accountability in AI development, open-source initiatives and collaborative research efforts can be used. Additionally, including patients and advocacy groups in the development and evaluation of AI systems can clarify the actual healthcare disparities in the real world and reflects on how the AI systems can meet the needs of heterogeneous patients. Part of responsible deployment and use of AI technologies is education and training of healthcare professionals on AI fairness and bias mitigation. With the diversity of datasets, better model interpretability and regular fair assessment, inclusion of bias minimizing approaches, and equitable healthcare outcomes for all of the patient groups can be achieved through AI development. Ultimately, fairness in AI medical application is not just about improving technical accuracy but also about the fact that AI driven healthcare solution must obey the principles of the equity, inclusiveness, and the patient-centered care.

3.2.1 Addressing Algorithmic Bias in AI Models

Machine learning has revolutionized many industries by automating decision-making processes and improving efficiency. In recent years, data science, machine learning, and artificial intelligence have become increasingly prevalent in various applications, transforming industries and everyday life. As these technologies become more integrated into the real world, one of the significant challenges in responsibly deploying machine learning models is mitigating bias(D. Satishkumar, 2024).

Types of Bias

Bias in machine learning refers to systematic errors introduced by algorithms or training data that lead to unfair or disproportionate predictions for specific groups or individuals. Such biases can arise due to historical imbalances in the training data, algorithm design, or data collection process. If left unchecked, biased AI models can perpetuate societal inequalities and cause real-world harm. Biases can be further categorized into explicit and implicit bias:

- Explicit bias refers to conscious prejudice held by individuals or groups based on stereotypes or beliefs about specific racial or ethnic groups. For example, an AI-powered customer support chatbot may be programmed with bias towards promoting products from manufacturers that offer higher commissions or incentives to the company.

- Implicit bias is a type of prejudice that people hold or express unintentionally and outside of their conscious control. These biases can have an unconscious impact on perceptions, assessments, and choices because they are often deeply ingrained in societal norms. The collection of data, the creation of algorithms, and the training of models can all unintentionally reflect unconscious bias.

For simplicity, we can categorize data bias into three buckets: data bias, algorithm bias, and user interaction bias. These categories are not mutually exclusive and are often intertwined.

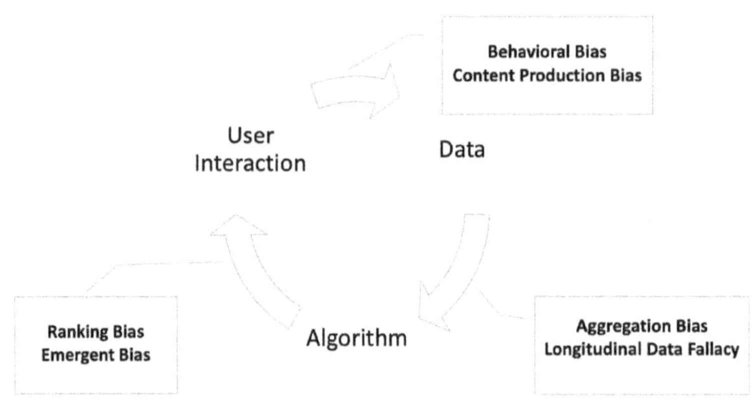

Figure 3.3: A Survey on Bias and Fairness in Machine Learning.

Source: (Mehrabi et al., 2024)

Bias in Algorithms

Bias in algorithms reveals the subtle influences of our society that sneak into technology. Algorithms are not inherently impartial; they can unknowingly carry biases from the data they learn from. It's important to recognize and address this bias to ensure that our digital tools are fair and inclusive. Let's look at the types of biases in algorithms:

- Algorithmic bias emerges from the design and decision-making process of the machine learning algorithm itself. Biased algorithms may favor certain groups or make unfair decisions, even if the training data is unbiased.

- User interaction bias can be introduced into the model through user interactions or feedback. If users exhibit biased behavior or provide biased feedback, the AI system might unintentionally learn and reinforce those biases in its responses.

- Popularity bias occurs when the AI favors popular options and disregards potentially superior alternatives that are less well-known.

- Emergent bias happens when the AI learns new biases while working and makes unfair decisions based on those biases, even if the data it processes is not inherently biased.

- Evaluation bias arises when we use biased criteria to measure the performance of AI. This can result in unfair outcomes and lead us to miss actual issues.

Bias in Data

When biased data is used to train ML algorithms, the outcomes of these models are likely to be biased as well. Let's look at the types of biases in data:

- Measurement bias occurs when the methods used to record data systematically deviate from the true values, resulting in consistent overestimation or underestimation. This distortion can arise due to equipment calibration errors, human subjectivity, or flawed procedures. Minor measurement biases can be managed with statistical adjustments, but significant biases can compromise data accuracy and skew outcomes, requiring cautious validation and re-calibration procedures.

- Omitted variable bias arises when pertinent variables that impact the relationship between the variables of interest are left out from the analysis. This can lead to spurious correlations or masked causations, misdirecting interpretations and decisions. Including relevant variables is vital to avoid inaccurate conclusions and to capture the delicate interactions that influence outcomes accurately.

- Aggregation bias occurs when data is combined at a higher level than necessary, masking underlying variations and patterns within subgroups. This can hide important insights and lead to inaccurate conclusions. Avoiding aggregation bias involves striking a balance between granularity and clarity, ensuring that insights drawn from aggregated data reflect the diversity of underlying elements.

- Sampling bias occurs when proper randomization is not used for data collection. Sampling bias can help optimize data collection and model training by focusing on relevant subsets, but may introduce inaccuracies in representing the entire population.

- Linking bias occurs when connections between variables are assumed without solid evidence or due to preconceived notions. This can lead to misguided conclusions. It's important to establish causal relationships through rigorous analysis and empirical evidence to ensure the reliability and validity of research findings.

- Labeling bias occurs when the data used to train and improve the model performance is labeled with incorrect or biased labels. Human annotators may inadvertently introduce their own biases when labeling data, which can be absorbed by the model during training. Corporate machine learning training courses are vital for ensuring high-quality, unbiased data, leading to more accurate and reliable machine learning models. Investing in such training empowers employees to enhance the overall effectiveness of machine learning initiatives.

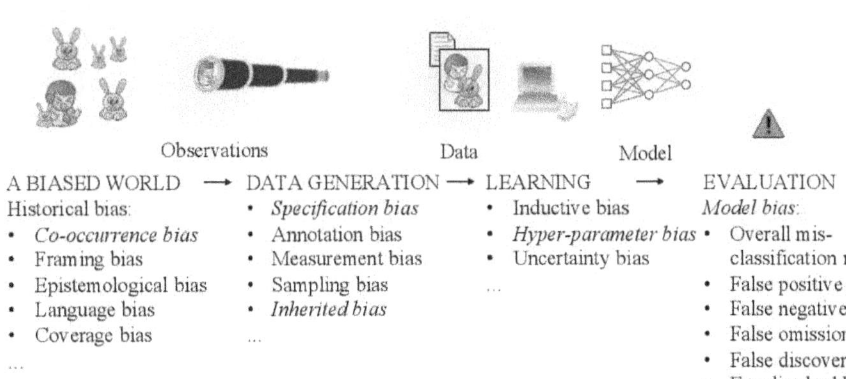

Figure 3.4: The Bias Pipeline in Machine Learning.

Source: (Schelter & Stoyanovich, 2022)

Bias in User Interaction

User biases can be introduced into the model through user interactions or feedback. If users exhibit biased behavior or provide biased feedback, the AI system might unintentionally learn and reinforce those biases in its responses.

- Historical bias is inherited from past embedded social or cultural inequalities. When historical data contains biases, AI models can perpetuate and even amplify these biases in their predictions.

- Population bias occurs when the AI prioritizes one group over others due to the data it has learned from. This can result in inaccurate predictions for underrepresented groups.

- Social bias can arise from cultural attitudes and prejudices in data, leading the AI to make biased predictions that reflect unfair societal views.

- Temporal bias occurs when the AI makes predictions that are only true for a certain time. Without considering changes over time, the results will be outdated.

Impacts of Bias in Healthcare

Training the AI systems on the healthcare with biased data results in misdiagnosis, inappropriate treatment recommendations, and less access to medical resources for different sections of the demographic groups. When we have bias in AI models– and to an extent, they are biased – in these AI models, they can disproportionately affect minority communities, and they can result in inequitable healthcare outcomes, and inequitable healthcare outcomes, in other words, reinforcing inequitable access to care, reinforcing inequitable quality of care. If the training data that goes into training an AI model is coming from a narrow set of people e.g. people of a certain race, ethnicity, gender, socioeconomic background then the model won't be able to discriminate or treat diseases in those groups of people. An example is if an AI system that has primarily been trained using data from patients with lighter skin tones does not detect dermatological conditions in patients of darker skin tones resulting in wrong treatment and delays. In the same vein, AI models used to assess the cardiovascular risk of Black and Hispanic patients, as is often the case, have been shown to underestimate the risks and leave them more likely to have undiagnosed or improperly treated heart disease.

AI healthcare systems can also be biased on the level of predictive models to be used for hospital resource allocation, patient triage, and treatment prioritization. In the case that these models are based on systemic inequalities to the available healthcare, the AI system might actually reinforce these disparities. For example, a model that predicts how patients should be triaged in an emergency room might rank patients according to historical data that, due to lack of an equal access to healthcare services, will disadvantage minority communities. Another common issue is gender bias: AI systems that are used to diagnose heart disease have been shown to work better for male patients than female patients based on the fact that heart attack symptoms are usually different in women, but the training data is usually male dominated. Moreover, AI systems based on socioeconomic factors (for example, location, insurance status, income) when making health outcomes predictions or prescription, can also define socioeconomic bias. In turn, such actions can negatively affect the quality of care given to people from disadvantaged backgrounds or those from remote populations. To address these biases, we need to work on a multi layered approach that will consist of diversifying the training datasets, using data augmentation techniques in improving the representation and also using bias detection and bias correction techniques on model development that can help in reducing these biases. Additionally, it is crucial to perform regular measurement and monitoring on the performance with varying patient groups to identify and override any but when developing AI healthcare systems to guarantee tractable consequences for any individual, even if they come from a distinctive history.

3.2.2 AI and Socioeconomic Disparities in Healthcare Access

Artificial Intelligence hold the potential to change health care by increasing diagnostic accuracy, reduction of treatment process and the quality of health care. Although there have been advancements in the implementation of AI and other technologies in healthcare systems, however, AI can make matters worse with less proper management. In most cases, access to AI based healthcare solutions are not evenly distributed, with wealthier population and well-funded healthcare institutions gaining more benefits from AI advances compared to economically disadvantaged communities. As the costs of developing, deploying, and maintaining AI systems are high, these systems are more feasible for large urban hospitals and specialized healthcare centers, but not for smaller healthcare facilities, rural hospitals, and clinics for low-income communities, lacking the financial resources and technical infrastructure, to adopt these. This leads to the fact that the AI driven diagnostic tools, treatment optimizing models, patient care platforms usually remain confined to the high-income areas, while underserved populations continue lacking the advanced healthcare solutions. This uneven access can perpetuate a divide between people with more and less access to the best quality healthcare, as those with privileged connections will be diagnosed and receive treatment faster, whereas population that are marginalized will be delayed when it comes to diagnosis and treatment(D. Satishkumar, 2024).

Data collected from such large medical institutions tend to serve higher income insured patients and often, AI models are trained on such data. That sets up a structural bias that is baked into the brain of AI systems: What is fed into them disproportionately consists of experiences that derive from the healthcare of more privileged communities, at the expense of low-income communities' patterns, challenges, and needs in health. For instance, an AI model trained on notes of patients from private hospitals might be better at diagnosing conditions frequently observed among prosperous patients but deteriorate and be incapable of identifying diseases prevalent among poor people, at least in these conditions' health of the low-income group is people. Additionally, hospital resource allocation, patient triage and treatment priority all use predictive models and such data reflects historical systemic inequities in healthcare access. A model that learns the historical difference of medical services to patients with low income may model these conditions and is likely to recommend fewer resources or lower priority of an equal patient in the future, thus further perpetuating existing disparities.

Other factors that may distort the system are insurance coverage and demographics. When using AI models to help manage patients in patient management systems, patient status (insurance, income level) may be taken into consideration and such models may recommend lower cost or lower quality treatment options for uninsured or underinsured patients. It makes this financial bias even worse and can further limit healthcare access of economically disadvantaged groups which only contributes to negative health outcome over time. Another example is AI based telemedicine and remote monitoring platforms with potential to extend reach of health care services to rural and underserved sectors that have dependence on reliable internet connectivity, digital literacy as well as smart device which may be lacking in low-income households. The effect is a digital divide: AI based healthcare solutions are hard to reach for the (very) people who could make the best use of them.

The addressing of socioeconomic disparities presents in AI driven healthcare calls for a full and highly targeted strategy. First, such datasets used for training AI need to be more inclusive and representative of the patient populations served by AI; that is, they would need to include data from underserved and economically disadvantaged communities. This will generalize AI models better in a way that they work for different demographic and socioeconomic groups. Second, we should first implement algorithmic fairness techniques, include reweighting data, adversarial debiasing and fairness aware training so as to reduce the biases in model predictions and recommendations, and secondly, we should AI developers and healthcare institutions should adopt algorithmic fairness techniques. Third, governments and healthcare organizations should subsidize access to AI enabled healthcare solutions in low income and rural areas through public private partnerships, infrastructure improvements and subsidized programs. Financial incentive and technical support to healthcare providers of underserved areas can accelerate the adoption of AI related technologies in healthcare and enhance healthcare delivery for those who are marginalized.

Furthermore, the AI driven healthcare systems designed for usage in low resource environments must be such to perform without any hitches. As an instance, such lightweight AI models would enable telemedicine as well as diagnostic platforms to be accessible to the communities in rural and remote areas. What makes this relevant is that AI solutions should also take into consideration the social determinants of health e.g. the state of housing, job, and access to a sufficient amount of nutritious food, to be able to provide more omnidirectional, context-aware healthcare recommendations. Additionally, including local communities and health care providers in the AI system design and evaluation can aid in discovering mismatch between care and gaps in care, and ensure that AI systems fit the practical demands of economically disenfranchised groups. Making sure that the development and deployment of AI in the healthcare systems occurs in a patient focused and equity centric approach will enable the utilization of the full potential of AI to improve health outcomes for all individuals irrespective of their socioeconomic background.

3.3 Regulatory and Compliance Challenges

The term regulatory and compliance challenges involve the obstacles and difficulties which organizations encounter in performing satisfactorily within a regulated industry with regard to legal, ethical and professional standards. These challenges are due to the necessity to follow certain laws, regulations and industry norms as prescribed by the government institutions and regulatory agencies. Regulatory challenges are to guarantee that everything in the processes, the way you handle the data, and the way you run the operation is in accordance with the legal frameworks like privacy laws and consumer protection regulations and other regulations of the industry. These standards must be met, otherwise the failure will result in heavy penalties — legal, fines, reputational damage, loss of customer trust and operational disruptions. The stakes are materially higher for some industries, like healthcare and finance, since the data concerns such as healthcare data and finance data would impact the lives of people(Kaufmann, 2017).

Compliance challenges are defined as a continuous and consistent adherence to standards and rules while being intransigent and transparent, accurate, and accountable. Organizations must apply good data protection measures, take the right business practice in line with industry regulations, and implement governance structure that will monitor and enforce compliance. Compliance means doing regular audits, training the employees on compliance requirements, and data breach and unauthorized access safeguards. Frequently changing regulations and having regulations across different regions, creates an increase in complexity of compliance on part of businesses and they need to be updated and made to adapt quickly. Compliance becomes even more complicated when these cross-border operations have to be accounted for as various countries and regions may have different requirements that would need to be complied with. For companies to safeguard transaction integrity, ensure regulation and comply with regulatory and compliance challenges, operational integrity and consumer trust must be maintained and legal and financial risks minimized. In addition, compliance also plays a role in creating a long-term stability and generating good reputation in the market.

3.3.1 Global Healthcare Regulations for AI Implementation

The regulatory environment for implementing Artificial Intelligence (AI) healthcare is complex and evolving on a national and regional level. Guidelines dictating how AI is adopted in medical diagnostics, treatment planning and care have been introduced by governments and the regulatory body to protect patient privacy, to ensure data security and to encourage increased clinical accuracy as deployed AI technologies continue to be integrated into these patient care solutions. Healthcare is about mainly sensitive data like medical histories, diagnostic reports and genetic information, which is why it is very important to set up strong regulations to avoid data breaches and misuse of data. To ensure patients retain the right and to prevent any harm in the AI driven healthcare systems, there is need to observe strict legal and ethical standards. These regulations are meant to not only secure patient data to avoid such disastrous and data privacy-enhancing consequences, but to also guarantee that AI based medical decisions are accurate, objective and ethically correct. These are regulations which must be complied with to gain trust of healthcare professionals and patients and also to comply with regulatory agencies approval(Sajja & Akerkar, 2016).

For instance, despite the efforts being made, there is no global harmonization in AI healthcare regulations, which poses a great challenge for developers and even healthcare institutions that operate in different markets. As AI evolves, organizations need to adapt to difference in legal frameworks, privacy laws and medical standards across countries in order to ensure AI model and compliance strategy can fulfill local requirement. As an example, once and aged an AI application has been approved to use under the Food and Drug Administration (FDA) in the United States, it may have to pass through the additional validation to meet the EMA or other regulatory bodies standards. This brings a extra layer of difficulty for AI developers and healthcare providers who are interested in deploying some components of AI solutions on a worldwide range. Along with that, the speed that technologies and advancements in AI are running has much faster than the creation of new regulatory frameworks, filling voids and inconsistencies of oversight. The challenges posed to these need to be solved collectively by international regulators, healthcare institutions, and developers of AI, to create a unified and transparent environment for the use of AI in healthcare. Here are some key points:

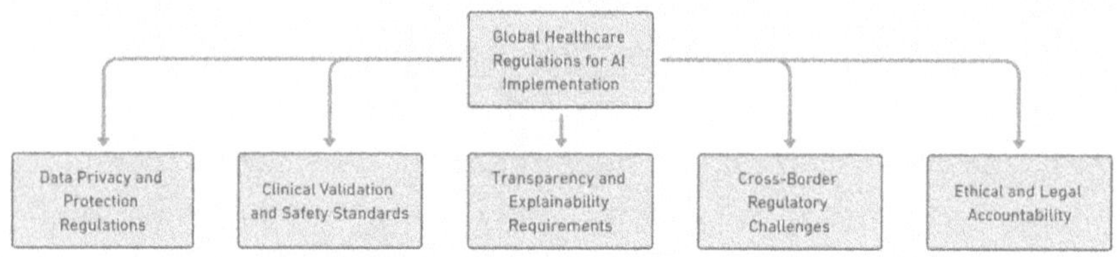

Figure 3.5: Global Healthcare AI Regulation Framework.

Source: (Self-generated)

1. **Data Privacy and Protection Regulations**

   Healthcare regulations worldwide prioritize patient data privacy, with the European Union's GDPR regulating data collection, processing, and disclosure. AI use requires informed consent, anonymization, and data breach prevention. The US HIPAA mandates confidentiality, integrity, and availability of protected health information. In Asia and South America, new regulations require AI developers to adapt to changing laws and develop flexible protocols.

2. **Clinical Validation and Safety Standards**

   Educated computer systems, such as AI driven medical devices and diagnostic tools must go through high clinical validation and safety standards before they are used in health care settings. In the US, AI based medical device is regulated by Food and Drug Administration (FDA). Testing for assessing AI systems recommending a diagnostic or treatment is equally important to test accuracy and reliability and safety of such systems. As with the AI based healthcare applications, the European Medicines Agency (EMA) monitors AI based healthcare Applications within the EU, and complies with medical device regulations and needs evidence of clinical effectiveness. Furthermore, countries such as Japan, Canada and Australia have approved their own processes for AI based medical technologies to meet local safety and performance standards.

3. **Transparency and Explain ability Requirements**

There is an increase in the regulators' focus on the transparency and explain ability of AI models employed in the healthcare domain. Such systems cannot be trusted in the absence of trust in the method by which they generate predictions, treatment recommendations and diagnoses — all of which must be understood by the healthcare providers and patients. For instance, the EU's AI Act includes particular recommendations that AI developers must explain the details about the data sources used, the model training process, and decision-making logic. In the U.S., the FDA promotes the use of interpretable AI models so health care professionals can validate and understand AI driven recommendations. Transparency makes sure that biases are averted, errors minimized, and ultimately increase the general acceptance of AI in health care.

4. **Cross-Border Regulatory Challenges**

Dealing with the regulatory wrangling of multiple countries also adds to the challenge of developing in AI for paid customers. Take, for example, a single AI-based diagnostic tool that has to be approved by FDA in the US, comply with GDPR in Europe and follow the regulations by the Pharmaceuticals and Medical Devices Agency (PMDA) in Japan. As a result, AI developers have to adapt their systems and documentation to satisfy local requirements that are complex. On top of that, data privacy standards differ internationally, clinical validation protocols have to be standardized and patient consent is subject to different requirements. The lack of global AI healthcare regulations will lead to different rules and regulations in various countries and regions, which causes regulatory friction to hamper the safe adoption of AI in healthcare.

5. **Ethical and Legal Accountability**

Ethical implications of AI in healthcare interests have been signposted by the regulators. Because AI model bias, lack of access to healthcare empowered with AI, and general automation's powerlessness to adjudicate matter of life and death need to be regulated and accordingly, have clear legal standards. The guidelines in the EU's AI Act, including those relating to fairness and preventing discrimination in AI based healthcare systems, will apply in Europe. Other national agencies including the FDA are also trying to put into place legal frameworks so that when AI driven recommendations cause a patient to suffer harm or a misdiagnosis there is some sort of liability. It will be important to establish clear guidelines on accountability and ethics practices as regards the use of AI to improve, rather than destroy, the quality and fairness of healthcare delivery.

3.3.2 Compliance with Data Protection Laws (HIPAA, GDPR)

Protecting patient sensitive information in healthcare mandates that care organizations ensure compliance with data protection laws. The reason behind the necessity for the implementation of strong data protection measures so advanced with the rise of digital health record and AI driven healthcare solutions, which are generating high volume and the high complexity of data being processed. Electronic health records (EHRs) handled by AI require a number of strict regulations to maintain patient confidentiality, not to mention preventing data breaches and keeping the healthcare organization in the legal zone. Noncompliance with these regulations can lead to severe consequences: likes of financial penalties, legal action and damaging of reputations, which makes compliance mandatory for both healthcare providers and developers of AI.

In the United States, there exist two of the most influential and commonly followed data protection regulation, which includes the Health Insurance Portability and Accountability Act (HIPAA) and in the

European Union is the General Data Protection Regulation (GDPR). HIPAA, as well as GDPR, set out very specific directions on how patient data can be collected, stored, processed and shared. These regulations safeguard patients' highly private health information and give patients more control over the use of their data. To do so, healthcare organizations and the AI developers must adopt security data management procedures such as encryption, access control, and audit trails, to care that the patient data is kept safe all times. Organizations also need to make sure patients know their data will be used in the way described and that they can choose to modify or delete it.

These regulations can have severe penalties in terms of financial, legal consequences as well as tarnishing an organization's reputation in the event of noncompliance. Furthermore, healthcare data is becoming more complex and the AI-based healthcare solutions are cross border. As AI developers work on integrating various bits of health information into the AI's algorithms, healthcare providers and AI developers must set clear protocols for how the data can be accessed, encrypted, anonymized — but the patients should still be the ones who control their own personal health information. Appropriate data protection also improves trust in AI-driven healthcare systems and protects a patient's privacy. Such a secure and transparent data handling framework helps building trust among patients and stakeholders to validating the credibility of AI-based healthcare solutions.

Figure 3.6: Healthcare Data Privacy: HIPAA and GDPR.

Source: (Mosaiyebzadeh et al., 2023)

HIPAA Compliance

Health Insurance Portability and Accountability Act (HIPAA) is passed in 1996 to protect the privacy and security of health information in the United States. It applies to health care providers, health plans, healthcare clearinghouses and any entity that handles protected health information (PHI). HIPAA lays down some mandatory standards regarding handling of patient data, including stringent standards to prevent data breaches and ensuring data integrity. However, healthcare organizations need to make sure that they follow HIPAA's privacy and security rules at all stages of data processing and handling, so they would not experience legal and financial penalties(Edemekong et al., 2025)also known as the Kennedy–Kassebaum Act or Kassebaum-Kennedy Act, comprises 5 Titles, as mentioned below. Title I: Protects health insurance coverage for workers and their families during job changes or losses. This Title restricts new healthcare plans from denying coverage based on preexisting conditions. Title II: Addresses healthcare fraud and abuse, implements medical liability reform, and promotes administrative simplification by establishing national standards for electronic healthcare transactions and national identifiers for providers, employers, and health insurance plans. Title III: Provides guidelines for pre-tax medical spending accounts and introduces changes to health insurance laws and deductions for medical insurance. Title IV: Offers guidelines for group healthcare plans, including modifications to health coverage provisions. Title V: Regulates company-owned life insurance policies, provides provisions for treating individuals without US citizenship, and repeals financial institution rules related to interest allocation. Questions to Consider Why was HIPAA established? The statute aims to establish confidentiality systems within healthcare facilities and beyond. The primary goal of HIPAA is to protect the privacy of PHI. Whom does HIPAA cover? All individuals working in healthcare facilities or private offices. Students. Non-patient care employees. Healthcare plans (eg, insurance companies.

HIPAA mandates and outlines the use of a broad range of security and privacy principles in order to protect patient data from unauthorized access and misuse. Patient data is required to be encrypted both in transit and at rest to prevent breaches under the legislation. HIPAA also mandates how patient data should be accessed and shared, with only the appropriate personnel being able to access patients' health information. Maintaining comprehensive data records of who accessed the data, when, and for what reason are part of any healthcare organization's duty to document trail.

Under HIPAA, healthcare organizations must:

- Data Encryption in Transit and at Rest: Data sent from your device to the server should be encrypted, as well as stored on the server. This is to guarantee that if the data is intercepted, it is not accessible without the correct decryption keys.

- Role Based Access: Sensitive health data should not be accessible to everyone, only the authorized personnel should be able to access it with proper role-based access protocols. To stop unauthorized access to organizations, multi factor, authentication and identity verification must be built in.

- Audit Trails and Monitoring: Accountability is ensured, and there is a way to find out if there was a breach, as systems need to be able to keep logs of who accessed the data, when the data was accessed, and why the data was accessed. The regular audits and real time monitoring will help to detect suspicious activities and also stop the data leak.

- Notification of the Affected Persons and Reporting of Breach: In the case of a data breach, organizations must immediately notify the affected persons, as well as other responsible bodies. Notifications must be issued within a specified period of time, and the organizations must specify what the nature of the breach and what steps they are taking to resolve it.

- Security Controls for Exposing Risk: Organizations only need to retain and dispose of data securely when it is no longer necessary to reduce exposure risk. HIPAA stipulates that data disposal must be done securely in order to avoid sensitive information recovery.

A HIPAA regulation violation can be punished by fines ranging from \$100 to as high as \$50,000 per violation and up to \$1.5 million per year. If the consequences of noncompliance are beyond financial penalties, healthcare providers as well as AI developers can see a reputation harm and loss of patient trust, with long term consequences. Additionally, HIPAA allows criminal charges for those responsible with imprisonment as a result of their repeated or intentional violations. To be HIPAA compliant, a total data protection strategy is devised which includes implementation of security protocols, staff training, and continuous monitoring of the practices.

GDPR Compliance

The GDPR was implemented in 2018 and applies to any organization that handles the personal data of European Union (EU) citizens, regardless of where the organization was located. GDPR adds very strict rules to the way patient data can be collected, processed and stored, placing responsibility on those who collect data to be completely transparent and ask for patient consent before their personal data can be used in any way. Healthcare providers in the EU who are not compliant with GDPR are not only required to do so, but also global healthcare organizations and individuals who provide services to EU citizens as well as AI developers who provide services globally.

GDPR also enforces several important principles such as data minimization, transparency and accountability. This dictates that healthcare organizations should collect patient data for legitimate and specified purposes and that patients know how their data is used. This calls back to the control patients should have over their personal information and reinforces this by giving them the right to access, modify, or delete data they have provided to the hospital at any time. To this end, the 'data protection by design and by default' needs to be implemented in healthcare organizations, which requires privacy and security features should be integrated in the AI systems from the start.

Under GDPR, healthcare providers and AI developers must:

- **Obtain Explicit Patient Consent:** When collecting or processing personal health data, organizations have to tell patients what for they are collecting that data and get consent. In order for patients to not be punished if they withdraw their consent at any time, patients should be given the choice to do so.

- **Patient Rights to Access and Control Data:** Patients have the right to have access to their data (the 'right to be forgotten' at any time. Procedures have to be devised by healthcare organizations to respond to patients' asks quickly and effectively.

- **Data Minimization:** The data should be collected only as required for the intended purpose and not to the extent that it is excessive. This minimizes the risk of data misuses and consequent exposure in case a breach occurs.

- **Data Protection by Design and by Default:** Security and privacy features have to come built in, from beginning, on AI systems and data handling processes. Included in that is encryption, anonymization, and secure data storage practices.

- **Breach Reporting and Response:** If a data breach is discovered, its discovery must be reported to regulatory authorities as well as those affected, within 72 hours. There has to be a clear incident response plan in place to minimize the damage and prevent such from happening again.

In case of non-compliance, GDPR imposes penalties of up to 4% of an organization's global annual revenue or €20 million, whichever is higher. There is also the possibility of financial penalties and the possibility of legal actions as well as reputational damage for GDPR violations. Despite the need to build healthcare's comprehensive data protection strategies that meet GDPR regulations, it is imperative that patient care delivery continues without too much disruption. When healthcare organizations process data across multiple jurisdictions, the complexity of GDPR compliance rises and also means that it becomes necessary for organizations to coordinate their compliance efforts.

When they follow HIPAA, GDPR and other data protection laws, healthcare organizations give patients trust that their data is secure, reduce risk of data breach, create safe environment for AI-driven healthcare solutions. To ensure compliance, staff need to be trained, infrastructure has to be secure, there has to be regular audits, and privacy of the patient has to be taken care of in a very serious manner.

3.4 Ethical Considerations in AI-Driven Decision-Making

The ethical implications involving patient care deemed fair and correct once AI is integrated into an ethical choice or decision within healthcare should be given ample attention and care. Since these AI systems can handle a lot of patient data and get patterns and generate clinical recommendations with the ability to be extremely fast and precise, they can out work human capacity in many cases. That ability to automate some of the most repetitive and tedious parts of medical care gives the potential to improve both the accuracy and the efficiency of a diagnosis and treatment. But the use of AI also urges the raising of serious ethics regarding fairness, accountability, transparency and patient autonomy. AI works based on algorithms and datasets compared to human decision making that is backed with years of medical training, clinical experience and standardized values of morality. However, the datasets also have biases, they embody historical inequalities, or don't take into account the diversity of the patient population. AI models trained on skewed or incomplete datasets can recommend things which will actually exacerbate the situation rather than resolve them, when these models are applied within the healthcare setting. This exposes the risk of becoming unbalanced and "clearly favoring for one patient group over another one and creating systemic inequities" by making decisions rooted in AI (Louise, 2011).

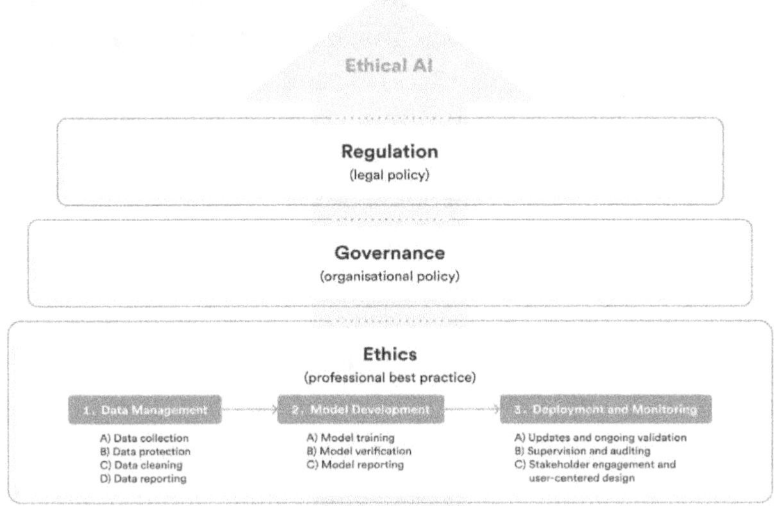

Figure 3.7: AI Ethics, Governance, and Regulation.

Source: (Daly et al., 2020)

The problem of transparency (or, to be more precise, the lack of transparency) is one of the most significant ethical challenges that come with AI generated recommendations. Most of the AI models are 'black box', which means even the developers creating them themselves may not know how the AI comes to certain conclusions. That's why asking clinicians and patients to give nod to AI generated recommendations without knowing how the decision was reached takes our confidence out of the system and brings up some very necessary questions about accountability. Determining who is responsible becomes difficult if an AI generated decision leads to harm or detrimental medical outcome. Who should be blamed if the healthcare provider that followed the AI's recommendation, the developer who created the algorithm, or the healthcare institution that implemented the system? In its absence, patients may feel powerless and helpless in the face of healthcare decisions powered by AI.

In order to address these concerns, AI based systems for healthcare must be designed in such a way that involves the adoption of strong ethical rules and transparent understanding of the decisions. Recommendations by AI systems should be accompanied by explanations that clinicians and patients can understand. The "explain ability" of this assures that as health care professionals, we can critically assess the outputted AI generated insights rather than blindly executing them. In order for AI developers, healthcare providers, and regulatory bodies to work in conjunction with one another to set protocols of transparency and accountability, there is collaboration needed. Patients should be given the right to know how AI was used in the care of patients, and healthcare organizations should put in place consent mechanisms so patients can opt to include AI in their care plan.

Additionally, it is important that human oversight be present to preserve the integrity of the AI used in healthcare decisions. AI should not replace human judgment and should be viewed as a support tool for the clinician. Clinical judgment or the patient's preferences should be overriding that of AI generated recommendations. Clear escalation protocols for case in which AI and human decision-making run counter to one another can prevent ethical breach and maintain patient care consonant with medical ethics. Detaching would need to be done continuously and audited critically to either uncover or avoid biases, errors, and unintended consequences. To do so, feedback loops need to be established with feedback to refine the AI models over time; these should reproduce the changes in the knowledge underlying medical care and patient diversity.

Another important aspect of ethical AI in the healthcare is the patient autonomy. In the use of AI in patients' treatment, patients should have a voice and be empowered to decide what is best for them. The clear communication between healthcare providers and patients on the role of AI in diagnosis and treatment is what is needed. Patients should be free to seek the opinion of human clinicians and to refuse to accept AI driven recommendations when they do not feel comfortable with them. Maintaining patient centered care and ethically sound care in the presence of AI based healthcare systems, is achieved by respecting patient autonomy but it also helps in boosting the trust in the AI based healthcare systems.

At last, healthcare institutions and AI developers have a social responsibility to prevent AI-based healthcare systems from exacerbating existing health disparities. It also means actively setting out things to remove any bias in training data, test and validate AI systems over different patient population, and ensuring equitable access to AI driven healthcare solutions. AI based healthcare technologies should not be left to be adopted only by high income communities, rural areas and underserved populations. In order to provide fair and effective healthcare outcomes for patients of all diversity, inclusive AI models are needed to develop that support the entire spectrum of human diversity.

3.4.1 Balancing AI Autonomy with Human Oversight

The ability to increase the use of AI in the making of healthcare decisions enables both positive and negative aspects of AI autonomy and human oversight. Medical data, the number of which is immense, can be processed by AI systems with speed and accuracy, and patterns found in the data can be used to make clinical recommendations. However, people need to avoid over reliance on AI and work in conjunction with it while maintaining proper human oversight, otherwise there are ethical and practical concerns associated with AI being used for excessive amounts. The crucial point is to maintain patient centered, evidence based, and ethics sound healthcare decisions on one hand while being able to maintain patient safety on the other hand with sufficient automation and human clinical judgment(Finnerty, 2024).

The Role of AI Autonomy in Healthcare Decision-Making

The AI systems are supposed to work autonomously to some extent with the help of the complex machine learning algorithms that will look for patterns in patient data and come up with medical recommendations. Nevertheless, this autonomy of AI raises a risk that AI made decisions can stake human clinical judgment or patient values.

Figure 3.8: Balancing AI Autonomy with Human Oversight.

Source: (Abayomi et al., 2024)

For instance, an AI system could suggest a cheap treatment without taking into account any patient preferences or at all fail to consider the complexity of the patient's medical history. In such cases, it is essential that AI recommendations are not blindly followed but rather that they are reviewed most carefully by healthcare professionals. It is essential to retain the authority given to the clinicians to ask questions, overrule or change AI driven decisions in cases of disagreement with clinical expertise or ethics.

- **Ensuring Human Oversight and Clinical Judgment**

  Human oversight is essential for limits putting up with and possible bias of AI models. They are trained on past data, which may reflect existing inequality and prejudices in health care delivery. Yet, if not managed, AI models may in fact reinforce a patient's demographic biases, and thus have unequal treatment outcomes. AI generated recommendation is a healthcare provider need to critically assess them and see if there are any inconsistencies or biases. For example, AI models may be more accurate for patients in the training data who are more common than rare and poor at subtle changes in disease caused by the difference in rare group in the training data seen compared to underrepresented people. Neither can these disparities be unnoticed without human oversight, which would perpetuate healthcare inequities.

Healthcare organizations who wish to integrate AI and human oversight must have clear guidelines for reviewing and approving AI generated recommendations. It includes defining the scope of AI autonomy, establishing the decision thresholds, and mechanism for clinician to challenge or replace AI based decisions. AI models should be used by clinicians only when they understand how they work, how they perform, and what they do well and not so well to understand when they should rely on AI and when they should use their own judgment. In addition to this, healthcare organizations should set up feedback loops for clinicians to report AI errors or inconsistencies to improve the model in continuous and increase its reliability.

- **Transparency and Explain ability in AI Recommendations**

It is important to have transparency in the AI decision making to fight for human oversight. Those using AI models to provide recommendations to healthcare providers and patients should be able to see how they work based on what data their model used, the algorithm they relied on, and the weighting factors they selected. Explanation ability also makes sure that the AI generated recommendations are not seen as arbitrary or opaque, instead being explained by clear and understandable logic. In fact, AI systems should also not be just giving a single rigid recommendation, but instead present multiple options of alternative treatment paths. By taking this approach, you incentivize the collaboration between AI systems and human clinicians on patient care decisions, particularly when the patients' care decisions have to be made based on both the human expertise and the advanced technological insights.

- **Establishing Ethical Guidelines and Oversight Committees**

Artificial intelligence autonomy must have boundaries on the frames of ethical governance frameworks. The AI performance should be monitored by independent ethics committees and regulatory bodies to discover how wrong it can go, to investigate adverse outcomes, and when the cases become complex to provide guidance. AI based decision should be regularly audited by health care institutions to identify patterns of bias, misdiagnosis or unintended harm. The roles and responsibilities of healthcare providers, AI developers and patients in the decision-making process should also be ethically stipulated. With proper structuring of the framework of AI autonomy to human oversight, healthcare organizations are able to keep AI driven decisions on track to the standards of medical, patient welfare and ethical principles.

3.4.2 AI in End-of-Life and Critical Care Decisions

Decision making in end of life and critical care are increasingly becoming sensitive and complex and often involve decisions that are of great importance. In such cases, medical teams must decide on how far to extend a patient's life while still trying to keep it as high a quality as possible. Utilizing these external data, AI has the potential to analyze patient data, predict outcomes and give recommendations for treatment plans. Nevertheless, these decisions are not purely clinical: In addition to being ties to clinical medicine there are personal, emotional and ethical issues involved. Consequently, AI driven recommendations must, in addition to medical evidence match patient preferences, values and dignity. The ethical dilemma is that AI should not take the groundwork of humanity and human compassion, and human autonomy, away from us in the moments that matter(Woods, 2006).

Often, decisions related to end-of-life care are complex trade-offs, which include life-threatening care versus comfort care. Large datasets can be analyzed by the AI systems to determine when curative treatments are no longer effective and providers can be suggested as a pain treatment or palliative care. Nevertheless, ethical dilemmas are encountered when AI recommends something that the patient wants or otherwise contradicts established guidelines or clinical. It is essential that human oversight and transparency are applied to

AI in order to prevent it from taking over patient autonomy. In these emotionally charged situations, AI should be used as a supportive tool of human judgment for healthcare providers to remain at the center of decision making.

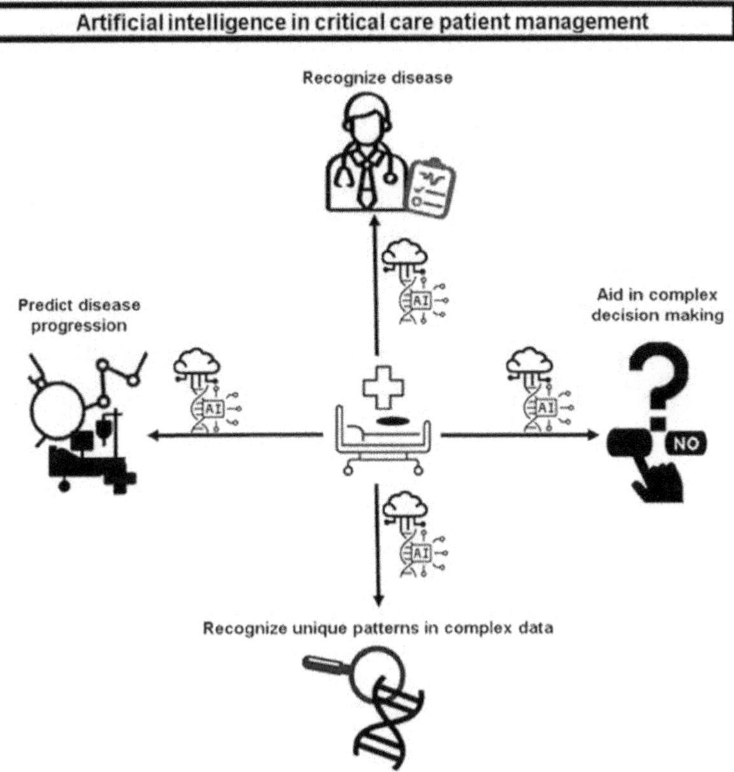

Figure 3.9: Artificial intelligence in critical care patient management.

Source: (Saqib et al., 2023)

1. **Supporting Palliative and Comfort Care:** Healthcare providers can use AI to predict when patients are moving into end of life so they are better served with palliative care rather than aggressive treatment. Such predictive algorithms can assess a patient's response to treatments, or the way one's disease progresses, or overall condition, to recommend comfort-based care when curative treatments are no longer achieving. For instance, the patient specific factors, like pain levels, mobility issues,

emotional well-being, can be considered by AI models to help customize palliative care plan. This allows healthcare teams to give more personalized, compassionate and patient's quality of life-oriented care.

2. **Facilitating Shared Decision-Making:** Trading life for quality of life is a common concept of end-of-life care decisions. They can share the decision–making process by offering real time, insight into risks and benefits of different treatment options. For instance, you could use AI to predict how much persons will live or if they can perhaps recover in different treatments, and this will help your patients and their families make the correct decisions. Concerningly, AI generated insights need to be in a transparent and understandable way to empower patients and their families to play an active part during the decision-making process.

3. **Respecting Patient Autonomy and Advance Directives:** Patient autonomy and its respect by AI systems and honor of advance directive like do not resuscitate (DNR) order or living will, are key to safe implementation of AI systems in healthcare delivery space. Electronic health records (EHRs) can store patient preferences that can be trained on, and recommendations can be made that align with them. There should also be mechanisms within ethical AI that flag inconsistencies between AI recommendations and patient directives, and these decisions should be reviewed and validated by human clinicians.

4. **Predicting and Managing Clinical Deterioration:** In critical care settings AI can determine when a patient will have a sudden clinical spiral down from reading real time data such as vitals, labs, etc. AI systems generate early warnings of what can become life-threatening situations to the provider allowing for more prompt intervention. Although predictive AI in critical care needs to be designed with the intent of not over-treating or overly intervening, especially when comfort care is the preferred course of action. Therefore, AI driven recommendations should be aligned with patient's care goals and clinical guidelines to

ensure that the interventions are medically appropriate and ethically sound.

5. **Emotional and Psychological Support:** AI's potential role in critical care should go beyond clinical decision making, as it should provide emotional and psychological support to patients and families too. Round-the-clock information and emotional support can be offered by AI based virtual assistants and the chatbots providing help to families to understand medical situation and coping strategies. In addition, AI is able to identify the emotional distress of patients and caregivers in order to help healthcare teams provide counseling or other emotional support. Combining AI with human care teams to provide the human as well as clinical aspects of end-of-life care.

6. **Ethical Governance and Oversight in Critical Cares:** Strict ethical oversight should be enforced with respect to AI recommendations in end-of-life and critical care scenarios. In instances where the decisions have life or death consequences, healthcare organizations should set up ethics committees to assess the use of AI generated recommendations. These committees should review the training data for AI models to confirm that it is representative of diverse patient populations and medical conditions. Additionally, AI driven decisions in critical care should be checked and validated by human clinicians to guarantee that the medical interventions will adhere to patient's preferences and ethical standards.

3.5 Trust and Reliability in AI Healthcare Systems

These are fundamental for successful adoption of AI in the healthcare. Existing AI systems must be understood to perform competently in clinical decision making and diagnostics within the healthcare realm, and healthcare providers and patients must feel assured that the systems will accurately and consistently operate, especially as AI systems become more integrated into clinical decision making, diagnostics, patient care and so on. Trust is a key driver of accuracy, therefore the AI models must have high sensitivity and specificity to minimize false positives and negative, since those have the potential to be devastating to patient health. Equally important is having consistency across different populations of patients in order for AI models to be trained on large, representative datasets to avoid biases and achieve equitable performance. Additional trust within patients and health care providers comes from rigorous via clinical trials and real-world testing for AI systems to reliably perform under various conditions. It is also important to build trust, and building trust is one of the critical parts, which are explain ability and transparency. How do AI models generate insights and recommendations to the healthcare professionals is required to be understood by them? By being able to create models that are and interpretable, have clear reasoning, traceable decision making and user-friendly report, clinicians are able to take confident use of AI driven insights for patient care.

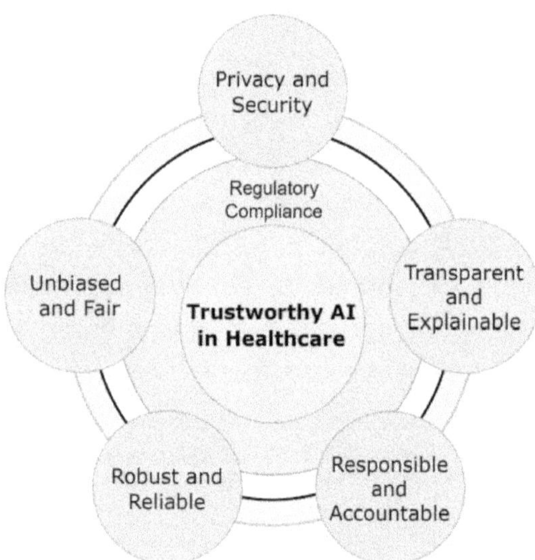

Figure 3.10: Framework of trustworthy AI in healthcare.

Source: (Albahri et al., 2023)

Sufficient data security and privacy form the basis for trust in AI healthcare systems. Thus, encryption and anonymization should be employed to protect patient data from breaches as well as unauthorized use and strict access controls should be employed for controlling access to dat. It also ensures compliance to healthcare regulations such as HIPAA, protecting patients and healthcare providers' data from getting misused and obeying to GDPR. Trust is built around another cornerstone such as addressing bias and fairness. Severe outcomes are unevenly distributed across different demographic groups; therefore, AI models should be trained on several available and diverse datasets to avoid that disparity in treatment outcomes. Bias audits and adaptive learning mechanisms that are performed on a regular basis are used to identify and address biases as they develop. Reliability can be maintained over time only through continuous monitoring and improvement. Feedback from the healthcare professionals should be included in the AI systems and their algorithms should be updated with latest clinical insights to enhance the accuracy and performance. The trust is reinforced with regulatory oversight from agencies such as the FDA and EMA making sure that the AI systems meet established medical standards that fall in line with other human dominated medical standards. Ultimately, the development of trust and reliability of AI healthcare systems is an ongoing process dependent on technological advances and alterations in the practice.

3.5.1 The Black Box Problem in AI Healthcare

AI healthcare black box problem is to explain how the output and recommendations of complex AI models, in particular deep learning models, are achieved. Most of the modern AI systems, including convolution neural network (CNN) for medical image analysis or natural language processing (NLP) model for stage clinical documentation are 'black box' – their inner decision-making process is opaque to the human users. Although these models are capable of producing very accurate results, the lack of transparency into the analysis, prevents parenting healthcare professionals from knowing how and why a particular diagnosis or treatment recommendation was made. However, this leads to a trust gap: clinicians don't want to trust the AI's insights without some explanation of what reasoning it is drawing on. For instance, an AI model can accurately predict the high risk of stroke of a patient but if the clinician cannot know which clinical factors led to that prediction, he may be reluctant to act on it. Accountability is also made difficult because it becomes difficult to identify and correct errors or bias when they occur, as there is little or no ability to explain AI decisions(Sendak et al., 2020).

In order to deal with black box problem, explain ability and interpretability are being emphasized in AI models. SHAP (Shapley Additive Explanations) and LIME (Local Interpretable Model-Agnostic Explanations) are techniques that are used to shed light to how an AI model weights the different inputs in deciding. For example, in a pneumonia diagnostic model, SHAP could show which factors (e.g. patient age, oxygen saturation, lung opacity on an X-ray) are the most important in the AI's prediction. Furthermore, in hybrid models, deep learning can be allied with rule-based systems or symbolic reasoning to add to interpretability in that it can place the outputs of AI mathematics in accordance with medical knowledge. There is growing demand from regulatory bodies like FDA and the European Medicines Agency (EMA) that AI systems run on healthcare use not only must be accurate, but also transparent in terms for how the decisions are made. An important way of bridging the trust gap in AI adoption in healthcare is the need to improve the interpretability of the AI models and offer actionable insights to clinicians as well as patients.

3.5.2 Strategies for Building Trust in AI-Driven Diagnosis

Trust building with diagnosis driven by AI requires an all-in accountability, accuracy, transparency, fairness and perpetual iteration. First, the main factor for establishing trust is accuracy and consistency. To make sure that AI models that are used for diagnosis are good for many patient demographics, medical conditions and clinical settings, the models have to be trained on a big plenty of data. However, for minimization of misdiagnoses and false alarms that undermine confidence in the recommendations of AI, high sensitivity and specificity are critical. Finally, real world clinical data or independent benchmarking of the model also validates the system and further builds trust by showing that the model's results satisfy the medical standard. Furthermore, the benefits of parallelizing separate AI algorithms via ensemble models can help to improve the diagnostic accuracy and thus reliability of the models. As an example of the rarity of results, deep learning models enhance the performance of AI systems for cancer detection like breast cancer in mammograms when they are used along with traditional image analysis techniques.

Transparency and explain ability need to be equal as it sets the trust. However, clinicians have to understand how an AI model produces diagnostic recommendations in order to truly and reliably incorporate it in patient care. This enables the clinicians to validate the reasoning of the AI by providing interpretable insights such as which medical features (e.g. lesion size, blood markers) had the most influence on the AI's decision. Use of techniques such as SHAP (Shapley Additive Explanations), LIME (Local Interpretable Model-Agnostic Explanations), can make a complex AI model more transparent by splitting it the contribution of several data points. The other key strategy is to ensure fairness and avoid bias. The way to avoid biases in diagnosis between patients due to the dataset used for training AI models and regular bias audits: train on diverse datasets. Finally, an ongoing monitoring and feedback loops allows for the process to always improve. If this accuracy rate of 98% is not the highest grade, AI systems should be updated regularly as a result of new medical research and clinical feedback to continuously improve diagnostic accuracy in accordance with the trend of health. Reinforcing trust in an AI driven diagnostic system, regulatory oversight by authorities of health for example, the FDA and EMA, ensure that the AI driven systems meet pre-established standards of safety and performance. Here are some key points:

- **Ensuring Accuracy and Consistency:** The first step to building trust in AI models for diagnosis is to ensure that the accuracy and consistency of AI models used to explain diagnosis reaches high levels. To enable reliable performance with a patient demographic, medical condition, or clinical setting and to avoid a failure at the delivery, AI models can't be trained on small data or too few, diverse data. Misdiagnoses and false alarms cause

AI recommendations to be undermined if they are insensitive and cannot be relied upon with the necessary high sensitivity and specificity. Further strengthening trust of AI systems came from validating with real life clinical data and independent benchmarks to prove that the model was performing at par with existing medical standards. Further, the combined properties of diversified ensemble models alleviate the accuracy and reliability of AI based diagnosis. A good example is that the AI systems for breast cancer detection on mammograms using conventional image analysis techniques have improved when deep learning models are overlaid on top of them.

- **Enhancing Transparency and Explain ability:** Transparency and explain ability is crucial in building the trust regarding AI driven diagnosis. Human clinicians need to be able to trust how AI models are arriving at diagnostic recommendations so we can pretty confidently bring them into patient care. This provides interpretable insights such as which medical features (e.g. lesion size, blood markers) influenced the AI's decision, which can be used to validate the reasonings of the AI by clinicians. SHAP (Shapley Additive Explanations) and LIME (Local Interpretable Model-Agnostic Explanations) are techniques, for example, they can make complex models AI turned out to be clearer by dismembering the demand of each piece of information. Because of the clear and understandable explanations provided by these AI systems, great levels of trust from the healthcare professionals and patients.

- **Ongoing Monitoring and Continuous Improvement:** There is no such thing as static trust in AI driven diagnosis, it needs to be monitored and improved constantly. Due to frequent changes in the medical research and clinical feedback, AI systems should be updated regularly and continuously to enhance diagnostic accuracy and accommodate with the changes of new health trends. Feedback loops involving clinicians reporting discrepancies and misdiagnoses on AI performance are established to continue improving AI performance over time. Trust is added to regulatory oversight provided by health authorities, such as the FDA and EMA, as the existence of AI driven diagnostic systems must adhere to predetermined safety and performance standards.

Multiple Choice Questions (MCQs)

1. **What is one of the primary methods used to secure Electronic Health Records (EHRs) in AI-enabled healthcare systems?**

   a. Data compression

   b. Blockchain encryption

   c. Image recognition

   d. Genetic algorithms

2. **Which of the following is a major cybersecurity risk in AI healthcare systems?**

   a. Increased data storage

   b. Unauthorized access and data breaches

   c. High processing speed

   d. Improved patient outcomes

3. **How can algorithmic bias in AI healthcare models be addressed?**

   a. Reducing dataset size

   b. Increasing model complexity

   c. Training on diverse and representative datasets

   d. Removing patient demographic data

4. **What is a potential consequence of socioeconomic disparities in AI-based healthcare systems?**

   a. Increased diagnostic accuracy

   b. Reduced healthcare access for marginalized groups

   c. Faster treatment outcomes

   d. Improved patient satisfaction

5. **Which regulatory framework governs data protection in healthcare systems within the European Union?**

    a. HIPAA

    b. FDA

    c. GDPR

    d. NHS

6. **What is the key ethical challenge in AI-driven end-of-life care decisions?**

    a. Ensuring data availability

    b. Balancing AI-driven recommendations with human values and patient wishes

    c. Increasing AI processing speed

    d. Reducing treatment costs

7. **What is a common reason for the black box problem in AI healthcare systems?**

    a. Lack of computing power

    b. Complexity of deep learning models

    c. Absence of medical guidelines

    d. Use of small datasets

8. **What technique can improve the transparency of AI-driven diagnosis?**

    a. Increasing hidden layers in the neural network

    b. LIME (Local Interpretable Model-Agnostic Explanations)

    c. Reducing training data

    d. Limiting model complexity

9. **What is an example of balancing AI autonomy with human oversight in healthcare?**

    a. Allowing AI to operate independently without review

    b. Combining AI recommendations with clinician judgment

    c. Replacing human doctors with AI systems

    d. Using AI only for administrative tasks

10. **What is a key strategy for building trust in AI-driven diagnosis?**

    a. Reducing transparency to protect intellectual property

    b. Regular updates based on new clinical data and feedback

    c. Removing human input from AI decision-making

    d. Limiting AI system access to patient data

Answers:

1	2	3	4	5	6	7	8	9	10
b	b	c	b	c	b	b	b	b	b

Chapter 4

Emerging AI and Healthcare Technologies

4.1 Integration of AI with IoT in Healthcare

Artificial Intelligence (AI) combined with Internet of Things (IoT) is involved in a real time monitoring, predictive analysis and automated decision making in the healthcare. Continuously, huge amount of health data is produced by IoT devices such as wearable fitness trackers, smart medical implants and remote patient monitoring systems. The data of this nature is processed by AI algorithms to tell patterns, detect the anomalies and predict the health-related problems before they become severe. This combination helps healthcare providers to able to provide individual care, increase the accuracy of diagnosis and improve patient outcomes. Early detection of health problems like irregular heart rhythms and the imbalances of glucose, with early intervention to reduce the number of emergency calls and readmitting people to the hospital. For example, intelligent smart insulin pump and intelligent pacemaker can adjust its functions automatically according to real time patient data, increasing accuracy, response as well as reducing the manual intervention needed(Shankar et al., 2021).

Healthcare now became more accessible and efficient with the help of AI and IoT in remote patient monitoring and telemedicine. Remote patient monitoring was enabled by the wearable devices or home-based monitoring systems that the healthcare providers use to track patient's health without too many in person visits. This data from these devices get inputted to AI based algorithms, which provide actionable insights to healthcare professionals, allowing them to make early diagnosis and also change treatment plan as soon as possible. For example, an AI glucometer tells how high a blood sugar is and recommends a change to the medication or diet. The use of AI and IoT enabled telemedicine platforms enables real time consultations for the doctors to provide better care even to patients in the remote areas. It cuts down healthcare costs by reducing hospital admissions and emergency visits whilst increasing the patient satisfaction through convenient and prompt care.

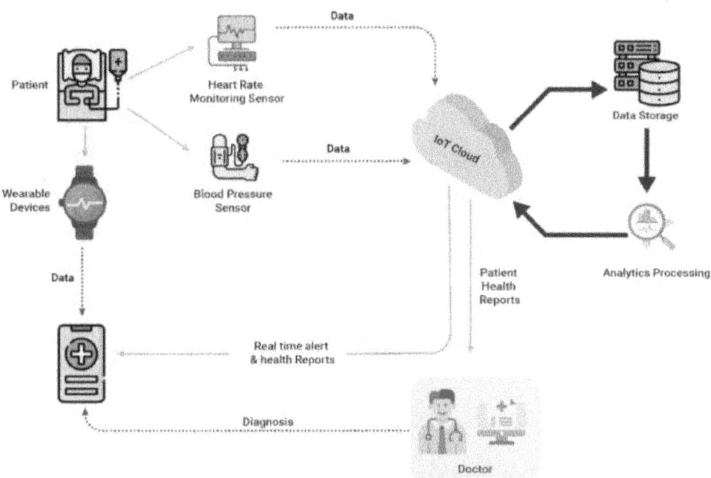

Figure 4.1: IoT-Enabled Remote Patient Monitoring System.

Source: (Boppana, 2019)

AI and IoT have also greatly improved the operational efficiency of health facilities. Medical equipment can be tracked by smart sensors, and medication inventory can be managed, as well as patient flow optimized, in hospitals. Data is analyzed by these systems to forecast patient admission rates, detect resource shortages, and therefore alter staffing levels to make use of resources efficiently. AI powered systems also provide better energy efficiency through controlling the lighting, heating and air conditioning based on the real time occupancy data to cut cost of operations while maintaining the comfort of patient. In spite of all of this, however, same goes for data security and patient privacy. Data security can be supported by AI through unusual access pattern detection and possible breach, while blockchain can secure, tamperproof data storage. Robust data protection measures ensure the trust in AI healthcare systems and help comply to the relevant regulations such as HIPAA and GDPR.

4.1.1 AI-Enabled Smart Medical Devices

Smart medical devices with AI aim at improving accuracy and responsiveness as well as efficiency of patient care. Smart insulin pumps, pacemakers, and automated drug delivery systems are developed to adjust their functions by real time patient data without direct involvement of healthcare providers. These devices contain AI algorithms that monitor continuous data streams of the patient's body and predict health changes, and respond automatically. For example, an AI powered insulin pump will monitor blood glucose levels and adjust the insulin dosage depending on the level, in order to avoid dangerous spikes or drops (Khang, 2024).

Figure 4.2: Medtronic Mini Med 670G system.

Likewise, pacemakers with AI can be adapted to changing heart rate and activity levels to achieve stable heart function under different physical and emotional condition. The patients' current health status can be calculated and they can get their required dose of the medication by using automated drug delivery systems.

Benefits of AI-Enabled Smart Medical Devices:

- Continuous Monitoring: Vital signs are constantly monitored by AI devices and there is instant feedback.

- Individual Patient Care: Results from individual patient data are used to adjust treatment.

- Health issues are detected early by AI that can reduce the complications.

- Smart devices do an automated drug delivery job, using those calculations to disperse exact medication dosages.

- Lowered Healthcare Costs: Cheaper hospital stays and fewer visits to emergency.

Smart medical devices functionality to provide real time adaptive response improves the quality of the patient care and reduces the need for human intervention. The devices capture lots of health data, and the AI algorithms are presented with all of it and will use the patterns it sees in order to detect anomalies, and later to forecast possible health complications. For instance, an AI enabled cardiac monitor can alert the patient's healthcare provider of irregular heart rhythms and inform the healthcare provider of the possible diagnosis and intervention. The ability to predict complications helps healthcare providers to anticipate the complications and change the treatment plan before it happens, improving the patient outcomes and reducing the emergency cases. In addition, AI enabled smart devices permit continuous remote monitoring of patients who suffer from chronic conditions such as diabetes and heart disease to enhance their control of their health in their homes. Not only does this increase patient comfort and convenience, but it also puts less stress on health care facilities through reducing the amount of hospital visits.

Operational Advantages of Smart Medical Devices:

- Better Patient Management: The health care providers can raise and monitor patients from elsewhere and can make instant modifications.

- AI Devices: AI devices get the patient's EHR into the synch for better decision making.

- Automating reduces the staff workload.

- AI predicts patient demand and allocates efficient resources in the hospitals.

- Data Driven Insights: Data driven insights assist in quick decision making for change in treatment strategies.

Smart medical devices with ai enable faster and more efficient medical device integration in creating better healthcare operations by increasing the data integration and communication between the healthcare systems. These devices can be plugged into the electronic health records (EHRs) and healthcare professionals can have access to the patient's data and history in real time. Integrated data is the outcome of applying AI algorithms that can analyze it and produce actionable insights for doctors that can be relied upon. In addition, AI driven automation frees up work from healthcare staff by doing routine monitoring and medication adjustments, freeing medical professionals to get it on complex cases and critical care. Also, the integration of AI enabled smart devices into the hospital infrastructure optimizes resource management through enhanced patient flow, reduced wait times among other parameters and reaches to maximum healthcare delivery efficiency. AI enabled smart medical devices are a very powerful tool for modern healthcare that combines the functionalities of real time data analysis, automated responses and the automation in the system integration.

4.1.2 AI in Emergency Response and Ambulance Coordination

AI is enabling the acceleration and improvement in medical interventions during emergency response and the coordination of ambulances. In high emergency situations where each second matters, emergency calls can be analyzed by AI driven systems for severity of the situation and suggestions of an optimal response. AI algorithms can handle large amounts of data from emergency hotlines, medical history of the patient, and current geographical information and find the nearest ambulance and outline the shortest route towards the patient. AI powered systems can also predict the patterns of traffic and road conditions and save emergency vehicles from the delays and bring the patients on a faster term. It ensures that patients in life threatening conditions like heart attacks, strokes or even severe injuries get timely medical attention and better survival rates and treatment.

However, real time clinical guidance and diagnostic support, provided by AI, also aid paramedics and first responders, during emergencies, as well. AI systems that are more advanced than what we already have integrated to ambulance equipment so the ambulance will be able to monitor a patient's vital signs of heart rate, blood pressure, and oxygen levels while on the way to the hospital. The real time data points can be analyzed by AI algorithms and they can detect critical signs of distress like irregular heart rhythms or respiratory issues. In some cases, AI can suggest specific treatments or change in medication before the patient arrives at the hospital to stabilize. An example of such an AI enabled defibrillator can automatically detect cardiac arrest, and automatically shock the heart to restore function. AI further creates a connection between the receiving hospital and the patient's hospital, sending patient details in real time and enabling emergency teams to render care to the patient upon their arrival.

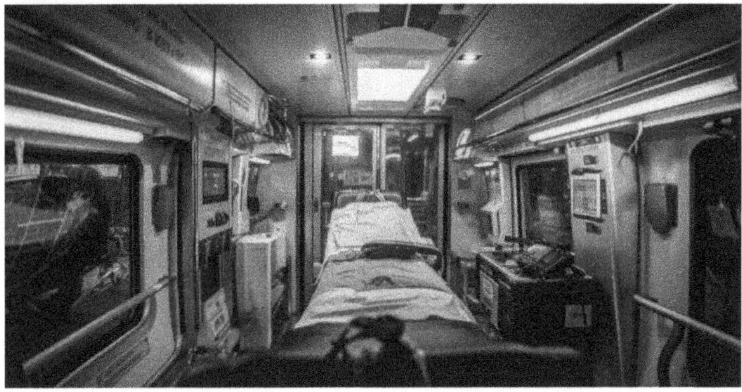

Figure 4.3: The Patient Compartment in ambulance.

Source: (Kibira et al., 2015)

Further, AI driven ambulance coordination systems minimize resource allocation as well as increase the efficiency of emergency response networks as a whole. With AI, the health authorities can predict the demand for the emergency services and so can pinpoint ambulances in areas where they will be required based on weather conditions and local events including, historical data. AI can continuously analyze the data of emergency calls and hospital capacities to propose the diversions to another hospital which are less crowded, making the treatment faster and cutting the patient waiting times. Also, AI can be connected to other emergency service such as fire and police department to enable communication with one another, especially when teams work in coordinating responses in such incidents as in case of multiple vehicle accident or natural disaster. Such coordination at this level is able to minimize response times, improve the patient outcome, and optimizes the operations of emergency healthcare systems.

4.2 Blockchain for Secure Medical Records

Medical records are becoming increasingly secured and patient data, its integrity and privacy, are being protected by the emergence of Blockchain technology. In healthcare industry, patient information is highly sensitive and is always vulnerable to unauthorized access, breach or tampering of data. Blockchain is a decentralized and immutable ledger system in which each transaction or data entry is recorded and linked to the previous one and thereby is almost impossible to remove or change data without detection. This makes sure that patient records are accurate and protect the data so that patients can have more power over them. By means of cryptographic keys, which are then only granted to specified healthcare providers, blockchain allows patients to grant access to their data to only those parties that are expected to have access. Also, blockchain is transparent, which means that it helps to build trust between patients and healthcare providers because all accesses and modifications of data are recorded and traceable (Udai Pratap Rao, Sweta Gupta, Piyush Kumar Shukla, Chandan Trivedi, 2021).

```
{
  "providedName": true,
  "name": "NoorulAain",
  "patient": "0x9965507d1a55bcC2695C58ba16F837d819B04Adc",
  "hospital": "0x3C44CdDdB6a900fa2b585dd299e03d12FA42938C",
  "admissionDate": {
    "BigNumber": {"value": "2022"}
  },
  "dischargeDate": {
    "BigNumber": {"value": "2023"}
  },
  "visitReason": "Migraine",
  "diagnosis": ["none", "migraine", "none", "migraine"],
  "patientID": {
    "BigNumber": {"value": "0"}
  },
  "allergies": "none",
  "geneticDisease": "migraine",
  "medicalReport": "migraine"
}
```

Figure 4.4: Structure of patient record on the blockchain.

Source: (UI et al., 2024)

It is one of the main benefits of blockchain for securing medical records as it prevents data breach and unauthorized access. Patient data stored in central databases, which can be hacked or single point failed. Whereas, on the other hand, blockchain deposits data on the network of nodes, so no cyberattack can do it. If just one node is compromised, the data is still secure on other nodes and any attempt to tamper with the data must be reached by the consensus of the majority of nodes in the network, which is impossible. Besides, the patient data is protected using advanced encryption techniques that keep it confidential even when intercepted. Another feature of blockchain is smart contracts that can actually carry data sharing and the process of controlling data access since only those doctors be able to access some data, based on a patient's consent.

Blockchain also helps healthcare institutions to streamline the process of sharing data with other healthcare institutions. Traditionally, patient records cannot be transferred from one hospital, clinic, or a specialist quickly and it is prone to errors, which delays the diagnosis and treatment. Blockchain allows for the centralized and secure platform for medical records to be accessed and updated in real time by authorized providers. Assume a patient visits an emergency room in a city different than the one he lives in, the attending physician can have immediate access to the patient's medical history, allergies and all medication records using a blockchain based system. This helps the patient get accurate and informed care as medication errors and adverse reactions are diminished. The health care system is being transformed by the blockchain technology, providing privacy and security of data while making all urgent information more accessible to patients.

4.2.1 *Enhancing Data Integrity with AI and Blockchain*

Healthcare data integrity is critical toward effective medical treatment and patient safety, and such data integrity should be maintained. In the healthcare industry, data integrity refers to the accuracy, consistency, and reliability of patient information, medical records, and treatment histories. Medical data errors or inconsistencies can cause misdiagnosis, waste in treatment, or worse, will be fatal. However, the huge amount of healthcare data captured by EHRs, wearable devices, and medical imaging systems poses a daunting challenge to healthcare providers to overcome and secure such information properly. Artificial intelligence (AI) combined with blockchain technology makes for a well-rounded solution in relating to its assurance of healthcare data that its accurate, consistent and secure.

Delivery of AI into healthcare has presented itself as the solution to processing large volumes of data quickly to find patterns, anomalies, and potential error in real time. Complex medical datasets can be sifted through by machine learning algorithms and abnormalities, as well as future health risk, can be predicted before they turn into problems. However, there is a system which protects this data from unauthorized access, tampering, or accidental data corruption. This security is offered by the blockchain technology that records each transaction or data entry in the form of an immutable ledger and links it to the previous entry using cryptographic hashing. Thus, it makes sure that after the data is recorded, it cannot be changed or deleted without the detection, so that there is the accuracy and reliability of the healthcare records. The combination of cybersecurity provided by blockchain, coupled with AI's analytical algorithms, creates a very safe platform to handle health data which health service providers can use to make more informed decisions and improve patient outcomes. What is more, blockchain's decentralized nature guarantees that data is still accessible to the right people, without allowing other parties to access the data or execute malicious attacks.

AI for Error Detection and Correction

It is essential for healthcare data error identification and correction with the help of AI. With Electronic Health Records (EHR) scanning, is can scan electronic health records to detect duplicate entries, mismatched test results, or wrong patient information. In addition, AI can also cross reference the data from different sources to verify the accuracy and consistency of the data. AI can also suggest the corrective actions when it detects an error so patient records stay up to date and accurate. For instance, an AI algorithm can detect inconsistencies in patients' medication history and sends the healthcare provider to fix so as to minimize chances of prescription errors. AI automation helps these error checking processes and, in turn, improves the overall accuracy in medical diagnosis and treatment plan to improve patient care outcome.

Blockchain for Secure and Transparent Data Management

It allows changes or updates in health data to be securely logged and verified among the network. Information is encrypted and linked to the previous entries so that it is practically impossible to alter or delete the information without being detected. The main difference of Blockchain in healthcare is the transparent and auditable trail of all data changes that makes the accountability and trust between the providers and patients. Blockchain can work with AI to analyze data, and ensure that any modifications of the data are legitimate and authorized. The combination improves the security and the accuracy of the data. This also means that blockchain is secure, and information shared between healthcare providers remains consistent, eliminating the chance of misdiagnosis or conflicting treatment plans as information could be variously. Through the intersection of the power to decipher and interpret data with the security and unchangeability of blockchain, healthcare providers can develop a more dependable and authentic data matrix.

4.2.2 AI in Decentralized Health Data Management

The shift from traditional centralized to distributed healthcare data storage models is referred to as decentralized health data management. In the traditional medical records in the healthcare system; data may be controlled, stored, and kept by hospital, insurance companies, or healthcare providers, resulting in data breached, lack of patient control and difficulties in online data exchange among the various providers. In this model, the healthcare data is stored on a blockchain based network, the data is encrypted, is distributed among all the nodes and is protected from unauthorized access. This model empowers the patients to have more control over their data as to whom should be able to view their records and under what circumstances. Artificial Intelligence (AI) simplifies the complexity of distributed data, maintain data integrity, facilitate secure access, and provide real time insights from a large amount of data in a decentralized health data management. AI and decentralized data systems integration enables patients to engage in better data security and to enhance the efficiency of healthcare delivery(Blobel, 2002).

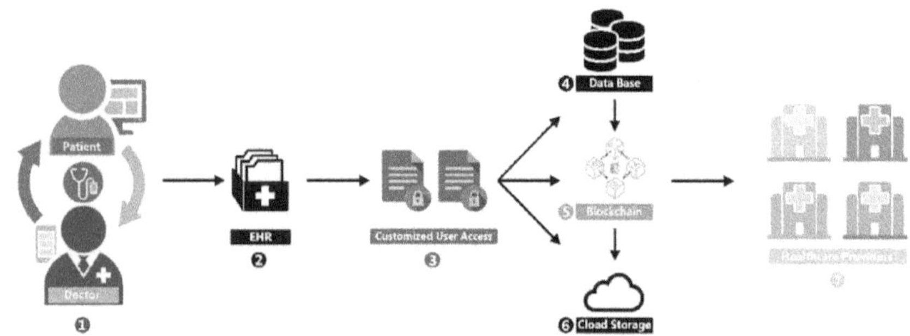

Figure 4.5: Healthcare data management in blockchain.

Source: (Khezr et al., 2019)

AI for Secure and Efficient Data Handling

A crucial role in maintaining safety and efficiency of decentralized health data systems is played by AI. Under a decentralized model, the patient data is spread across several nodes, which doesn't exist in a single central database. This complex structure is handled and processed by using AI algorithms. For instance, AI can examine patterns of data access to detect suspicious behavior, for example, incident attempts to access patient records. AI can also be used to trigger automated responses in case of breach or anomaly in order to prevent further unauthorized access. Similarly, data consistency across the network is also improved by the use of AI to verify and synchronize the records stored on different nodes. This is done so that the data of patients is accurate even from different healthcare facilities or systems. Additionally, AI can speed up the data retrieval process by improving the data indexing and retrieval within the decentralized framework, resulting in better response time of healthcare systems.

Meanwhile, AI also toughens the security of the decentralized systems by integrating with the blockchain based encryption and authentication mechanisms. Blockchain technology makes sure each data entry is encrypted and chained to the previous one thereby making it tamper proof. The blockchain activity can be monitored by AI and its ability to detect any discrepancies or unauthorized modifications can be used. For instance, if an unauthorized party tries to change patient data, AI will notice the inconsistency and correct it by restoring the original data. When combined with AI, it enhances the security of the integrity of healthcare data and reduces the probability of data breaches, and the confidentiality of patient sensitive information.

Patient Empowerment and Data Ownership

Imagining decentralization in health data systems means patient empowerment in which the patients can have greater control over their medical records. Patient data in centralized healthcare systems can be controlled by healthcare institutions who may have limited access to patient data. In decentralized models, the control is shifted to the patient and they get to determine where the access permissions live with blockchain based smart contracts. This is made possible by AI, which handles the right to access and data is shared only with authorized parties. For instance, a patient can give his primary care physician the right to see his medical history but deny such access to other healthcare providers not needed. Moving to a post-certification mode of operation for ITCC raises the question of what to do with all those nodes for which we are no longer ensuring that their SET version control procedures meet some certification standard.

AI also helps in the accuracy and completeness of patient data in decentralized systems. Learning from machines can locate generalized problems of patient records and advise healthcare providers to fix them. To illustrate, AI could cross reference data from other nodes in the network for a patient in question and complete the patient's medication history when it is incomplete. This helps healthcare providers have the whole and honest medical history of the patient so that they can make a better and informed decision. With the ability for patients to directly access and manage their data, trust is built and patient engagement is improved in this patient's own healthcare management.

Improving Interoperability and Data Sharing

Healthcare presents a big interoperability challenge, as various healthcare providers tend to use incompatible data storage systems for storing data that cannot be shared. The management of decentralized data through AI is making the health data interoperable by real time data synchronization and standardization. The data from various sources can be translated into a single format using AI algorithms so that patient records can be understood by all healthcare providers, regardless of their system. Regularly, AI can harmonize patient data received at multiple hospitals utilizing various Electronic Health Record (EHR) systems to provide a unified view of patient medical history.

Smart contracts also help removing the need for intermediate parties to process data sharing without prior consent. AI makes sure that patients define the conditions under which their data can be shared and then follow them. With a patient's consent, AI can be used to monitor the access and revoke permission automatically when the time limit to share the medical history expires. It helps further secure shared data and maintains the control of their medical information by the patient. AI increases the interoperability and reduces the sharing of data, making it possible for providers to care more coordinated, with less errors in treating the patients and providing better results.

4.3 3D Printing in Prosthetics and Implants

The 3D printing technology has played a leading role in radically changing the prosthetics and medical implants to a higher level of customization, precision and efficiency. The big difference with 3D printing as compared to traditional manufacturing techniques, in which components are cut, molded or assembled, is that it is based on digital models and manufactures objects layer upon layer. It enables the making of sophisticated and patient specific medical devices which can be designed to patients' specific anatomy and medical needs. 3D printing allows unique artificial limbs, hands, and body parts that fit like a natural experience of the functionality better than other mass-produced options. Much like medical implants like knee replacements, dental implants, as well as cranial plates, custom design of these can harmonize with a patient's body geometry for both comfort and clinical outcomes. In this, artificial intelligence (AI) takes the process of 3D printing one step further by analyzing patient data – MRI and CT scans, for example – to create the most precise designs possible and preform the prosthetic or implant under ideal and other parameter conditions. The intertwined 3D printing and AI will enable the healthcare providers to provide their patients with personalized treatment, thereby decreasing the risk of complications, and increasing overall patient satisfaction.

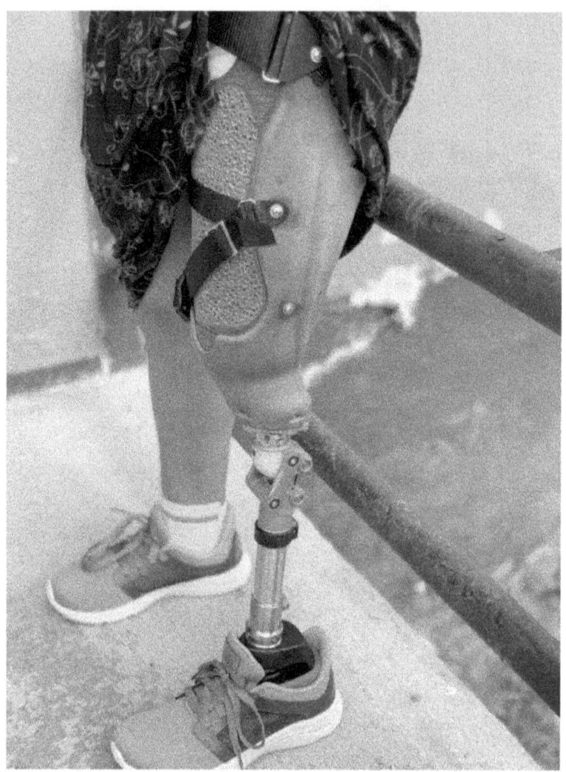

Figure 4.6: Prosthetic leg featuring custom 3D-printed prosthetic sockets with flexible inner liners.

Source: (Gutierrez, 2023)

3D printing is beneficial for prosthetics and implants in more ways than accurate, personalized fit. The major advantage is the decrease in production time and costs. 3D printing is relatively cheap and simplifies the production process by automatic Alize it and waste less material with high accuracy as the traditional prosthetics and implants. AI driven design optimization simulates the prosthetic or implant under stress and pressure, cutting waste and production time, as far as the prosthetic or implant is perfect, there is no need to adjust anything except for scheduling production. As a result, a lightweight prosthetic limb can be a 3D printed to equitably distribute weight, increasing comfort and mobility for the patient. By using AI and 3D printing, porous structures meant for better bone integration can be created for medical implants. For instance, the 3D printed hip replacement can be designed with a textured surface, that encourages the surrounding bone tissue to grow inside the implant, stabilizing it and reducing the possibility of rejection or failure of this kind of implant. The level of precision and functionality results in better clinical outcomes and faster recovery of the patients.

Another benefit of 3D printing is improved impetus to high quality prosthetics and implants, especially in remote or underserved areas. The current medical devices are manufactured in centralized facilities, and thus involve complex logistical efforts in order to transport and deliver the manufactured products. However, unlike that, 3D printing allows healthcare providers to produce prosthetics and implants on site or in the nearby medical centers, thus reducing the costs and time required for production. Remote design systems with AI are powered by remote design system that allows healthcare professionals to collect the patient data (body measurements and images) and automatically generates the customized designs that can be printed locally. In rural areas or low-income regions, there is limited specialized medical care and this helps healthcare access for patients. Moreover, 3D printing provides rapid prototyping support, which enables healthcare providers to try it as well as to modify the prosthetic design based on patient opinions within very little time. Through this iterative process, the final product will be what the patient needs, and thus comfort, functionality and patient satisfaction will be improved. Given this, it is possible that in the future bioprinter tissues and organs will advance further, lowering the degree of reliance on donor transplants and reinforcing better living for patients.

4.3.1 AI-Optimized Prosthetic Designs

The field of prosthetics has been revolutionized by AI optimized prosthetic designs such as highly customized and functional artificial limb. Focusing on traditional methodology of prosthetic while using manually adjustment and standardized template leads to discomfort and very little mobility for the patient. This process has been changed by AI which analyses patient specific data including body measures, gait pattens, muscle strength and medical imaging to create the tailor-made adaptable design with a natural fit and improved functionality. Because the prosthetic can be run through machine learning algorithms that simulate how the prosthetic will respond to stress, pressure, and movement, it can be adjusted in real time before production. AI driven design tools further allow for the creation of lightweight yet strong and flexible material, therefore reducing the weight of the prosthetic. AI provides a way to design a lower limb prosthetic with reinforced areas to endure higher impact while running or walking resulting into increased comfort and performance. Rapid prototyping and faster production mean cheaper and hence more accessible patient treatment. The ability to achieve such degree of flexibility and precision in making AI optimized prosthetics improves patient's comfort, mobility, and also enhances patient's quality of life, thereby becoming a breakthrough in personalized healthcare(Nawaf Mansour Saeed AlQahtani, Saad Mohammed Abdulaziz Alsaaran., 2023).

How AI Enhances Prosthetic Functionality

Prosthetics had been transformed by AI into a dynamic and adaptable area offering real time responses to the user's intentions and surroundings with the adaptability to the user's movements and postures. While traditional prosthetics do just mechanical functions, AI prosthetics enrich prosthetics with complex data processing and machine learning to achieve highly responsive and intuitive device. The ability to adapt and learn the unique way a user moves is by far the biggest advance in prosthetics of any kind. AI driven prosthetics are able to modify their response to the inherent data such as the gait and other biometrics, by analyzing the muscle signals. An example of this might be an AI powered prosthetic leg that adjusts its moves according to how an individual with the prosthetic might shift their body when switching from an even to uneven terrain. This predictive capability is also extended to myoelectric prosthetic hand, that is able to predict the user's intention regarding the gripping, pinching and opening of a hand based on the subtle muscle activity, thereby reducing the cognitive load and making prosthetic feel like a natural part of the body. What makes AI good at its job come to life in prosthetics is its fine ability to analyze and interpret complex data coming from muscles, nerves and even the brain to point out tasks such as grasping delicate objects or turning a key. This adaptability and precision assists in the functionality and ease of AI prosthetics, especially for users who lead an active life with traversing in some different natural components.

Figure 4.7: AI enabled prosthetic hand.

Source: (Saikia et al., 2016)

More particularly, AI powered prosthetics also perform extremely well in terms of adaptive responses to changing conditions, and continuous learning. With this, these devices are capable of adjusting the thrust in response to changes in the user's environment, if there is a need to transition from one surface that is very smooth to gravel or any other strange terrain, these devices do not need manual input to maintain balance or stability. This adaptation enables their use of the surroundings more confidently, and naturally. Additionally, AI driven prosthetics continuously improve their performance over time, based on how a user behaves and it gives feedback about that. While static traditional prosthetics have constrained choices in terms of how the prosthesis can be held or the force by which the prosthesis can be gripped, AI models can learn to better how a user would like to hold a particular object or to change grip strength throughout time, leading to a better overall user experience. AI has advanced prosthetics — multi articulating hands being just one example — which greatly benefits from the ability to coordinate multiple joints and components in real time so that users can tackle the complex task of sewing or typing effortlessly. User friendly interfaces allow adjusting to your needs and preferences with connected apps and this level of dexterity and customization is further supported. Continuous learning capability makes the prosthetic learn with the user, which increases trust and satisfaction.

4.3.2 AI in Bioprinting for Tissue Engineering

Major technological and materials advancements, utilized effectively as a part of lately created techniques and developments, have lifted AI to the upper positions in the field of bioprinting, permitting the precision, efficiency, and scalability of manufacture of intricate biological structures. Bioprinting is the process of layer-by-layer deposition of living cells, growth factors, and biomaterials to construct tissues with similar structure and function of the natural human tissue. Based on traditional bioprinting methods, structural integrity, cell viability and architecturally a complex tissue were often a problem. Bioprinting is addressed by AI by enhancing the bio printing process, designing and optimizing, through the analysis and prediction of data. Biological samples of large datasets may be analyzed by machine learning algorithms to create optimized printing patterns, predict cell behavior and optimize content in bionics to minimize incompatibility and provide better survival rates. Such modeling ability allows researchers to develop more accurate tissue constructs that when printed are functional in the way they were meant to be functional from the moment they are implanted in the body. Aside from simple structures, bioprinting is no longer a simple task by integrating AI as it is now possible to have highly vascular tissues, functional skin grafts, and even mini organs with higher success rates.

Monitoring and adjusting the printing parameters are the key features of AI, which can enhance process of bioprinting by turning out more accurate and structural integrity. Bio inks printed with AI driven systems are viscous enough to be fluid at the printing temperature, but not so viscous that they extrude during printing, while the cells in them are dense enough to fill the cellular space of the extruder channels, but not so dense as to reduce oxygenation during long print times. For instance, AI can foul out issues like uneven layering or death of cells (also known as failure of manufacturing) that take place during printing process and correct them in real time to minimize the material waste and increase the success rate of manufacturing printed tissues. In addition, computerized imaging and analysis of the printed tissue can judge the morphology and the activity of the cells and feedback for future printing runs. It also allows researchers to apply their designs for the printing of tissue to test the assumption and develop the design itself in advance of printing using AI to simulate biological processes. By optimizing this process in real time, the printed tissue suites' mechanical properties and biological functionality are wounds healed, drug testing and organs transplantation.

In addition, AI is applied in post printing analysis, as well as long term functionality of the tissue engineered product. Knowing whether the cells are growing and integrating into the host tissue, and doing so in a timely manner, ways that an AI algorithm could monitor can alert people early after there are some signs of rejection or poor cell viability. Machine learning models can forecast how printed tissue will behave under influence of environmental factors including blood flow, oxygen levels and mechanical stress and thus help researchers improve the tissue composition and structure. It also helps to develop complex multicellular tissues by studying how distinct cell types collaborate and determine where to place them during printing. This has enabled the creation of vascularized tissues of vascular structures which adequately support complex and large tissue structures. Simply, AI based bioprinting has the ability to enable more precise control to grow tissue of composition and giving rise to new regenerative medicine solutions for organ repair, transplantation and disease modeling. AI and bioprinting are combining to create fully functional patient specific tissues and organs — a major advance in the road to personalized medicine.

4.4 AI-Driven Virtual Assistants for Patient Support

Virtual assistances driven by AI have proven to be boon for the healthcare system in terms of patient support in the process through the assistance provided 24x7, as well as communication and streamlining patient care. These natural language processing (NLP) and machine leaning powered virtual assistants are able to understand

and respond to patient inquires live. AI driven assistant interface can be accessed via mobile apps, websites, smart devices, and patients can interact with them to get immediate responses to symptoms, medications and treatment plan. Whereas virtual assistant is especially useful for handling routine queries such as scheduling appointment, refilling prescription, or giving out post treatment care instructions, thus reducing the administrative burden on healthcare staff. Also, AI driven assistants can triage a patient's symptoms to the right level of care according to the urgency of the condition. This creates a capability which increases patient satisfaction by enabling timely and accurate support while allowing healthcare professionals to concentrate on cases that require human intervention (Gerdin, 2021).

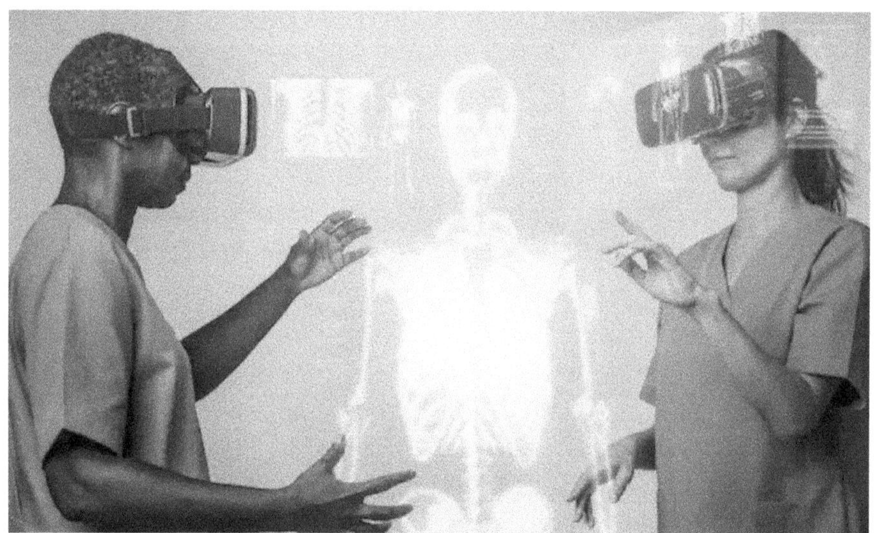

Figure 4.8: Virtual Medical Assistant.

Source: (Krishna et al., 2022)

But both patient education and chronic disease management made use of AI driven virtual assistants. These systems could provide personalized health recommendations to the patients that are related to the patients' medical history, lifestyle and existing treatments. An example is an AI assistant which tells a diabetic patient to take their blood sugar level check, recommends dietary changes and offers guidance to symptoms management. AI is able to analyze patient data over time and based on this virtual assistant can adapt and refine their recommendation, making them more and more relevant and effective. On top of that, AI assistant can be a source of mental health support with relaxation techniques, CBT exercises, and crisis intervention guidance given. This constant support provides the patients with the means to play an active part in their health and decrease the likelihood of emergency situations. Virtual assistants can also integrate their data with electronic health records such that healthcare providers can keep track of a patient's progress and modify treatment plans.

Moreover, AI enabled virtual assistants have the ability to enhance operational efficiency in the healthcare systems. These assistants automate the administrative tasks and manage the patient–provider communication in order to reduce wait times and increase patient access to care. For example, when we have an appointment scheduling conflict, AI can handle the conflict, automatically send the reminder, and follow up with the patients after the medical procedure to ensure a high compliance in the treatment plan. Virtual assistants driven by AI can also assist healthcare providers with access to patient data, clinical guidelines as well as findings of research in real time during consultations, helping to improve the accuracy of the diagnosis and treating. In case of emergencies, AI assistants can aid in quick communication among medical teams by expediting their work through the deliverance of critical information on the patient's past and current symptoms. This comprehensive support framework not only helps to provide a better experience for the patient but also reduces healthcare costs by reducing the dependency on visits which are not necessary and improve the overall quality of care. The application of AI-driven virtual assistants into the health system represents a transformation of care from offering more responsive and patient centric experience, where technology augments the scope of ability of humans to create better health outcomes.

4.4.1 AI Chatbots for Mental Health and Well-Being

Today, people who are dealing mentally and psychologically find AI chatbots a very useful resource for managing their mental health and well-being by providing free, instant and confidential service. These chatbots list their power in the fact that they use natural language processing (NLP) and machine learning (ML) algorithms to sound human like, understand human emotional cues and even respond personally. AI chatbots can offer techniques of cognitive behavioral therapy (CBT), mindfulness exercises, tips on how to deal with stress, etc. in the cases where a person feels anxiety, depression, and so on, 24/7 availability means that it's always available and available at any time of the day or night even if you don't have time or ability to visit a clinic to get some support. For example, an AI chatbot that for example is able to help someone who is going through an anxiety attack and can follow breathing exercises, or grounding techniques to get the stress under control. However, above all, these chatbots have the ability to learn to make suggestions as they learn about the user's pattern of thoughts and mental health needs with the help of AI over time to provide more personalized support.

AI chatbots are also used to screen for and prevent mental health problems at an early stage, apart from offering emotional support. If a person's conversation involves language, tone and sentiment, AI can analyze that and flag when a person's talk becomes more negative about themselves, shows signs of hopelessness, or if a person says they want to commit suicide. If these patterns are seen, the chatbot can the escalate situation by suggesting the user to check with a mental health professional or even directly connect the user to crisis support services. Dealing with these early is a proactive way of intervening and may prevent crises from even occurring. Besides this, AI chatbots are able to analyze the patterns of user behavior and their emotional states with the long term to assist mental health professionals in improving their treatment plans. Chatbots give anonymity and confidentiality to open up about sensitive issue, making the interaction more honest and supportive. More and more, mental health chatbots are getting better at being empathy, and are also filling a gap between mental health services and those who most need it.

4.4.2 AI in Personalized Patient Engagement

Personalization has been one of the key transforming forces in personalized patient engagement by allowing the healthcare providers to communicate, recommend, and care in a manner that is tailored to each individual patient's data. By performing advanced data analysis and machine learning using electronic health records (EHR) data and patient reported data, wearables data in addition to the variety of data points available from each patient, AI systems can create a detailed profile for each patient. This gives healthcare providers the opportunity to acquire patient's preferences, history of health and behavioral patterns, thus allowing healthcare providers to create plans commensurate to their needs. An example is an AI driven system that can oversee a patient's medication compliance, life style dispositions and past treatment results to make an exclusively customized changes to their treatment plan. In addition to giving treatments targeting to the patients' unique health goals and medical conditions, AI can also deliver personalized health tips, reminders of the medication and appointment, and diet and exercise suggestions. By communicating on a personalized level through AI powered platform, patients are better aware that they aren't being ignored and are supported which in turn leads to better adhesion as well as better outcomes(Tailor, 2015).

Predictive care—predicting the stages of a disease—along with proactive health management is included in the use of AI for patient engagement. The AI keeps track of real time collected data from wearable devices and remote monitoring system and translates it into the early sign of health deterioration which can be used for timely intervention. To give an example, if there is a patient diagnosed with hypertension which is being monitored by an AI system, this system is able to detect an increasing blood pressure and alert by recommending lifestyle changes and or check in with doctor. This engagement is enhanced by AI chatbots and virtual assistants that keep the patients engaged and answer their questions as well as steer them towards their health journey. This continuous interaction not only increases the responsibility of patients in their health but also gives healthcare providers the ability to foresee the potential problems before they escalate. Additionally, healthcare providers can determine specific wellness program measures and health campaigns to serve to various patient populations based on shared characteristics and risk factors enabled by AI systems. Personalization in patient care based on the AI will only become more sophisticated as we've crossed the threshold with AI systems.

4.5 AI in Mental Health and Wellness

Mental health care is trending with the revolution of AI and it is being applied by it through scalable accessible, and personalized solutions. It is common that traditional mental health services do not always cater to the needs of patients as there is limited availability of professionals to consult with often taking months, even years, for tests to be returned and reports to be released. Chatbots, virtual therapists, diagnostic tools etc. that function with the help of AI are filling in the gaps by providing immediate and round the clock solutions. However, mental health platforms are powered by AI and they act to analyze data from patients' interactions like that of a conversation pattern, mood indicators, and even data from wearable devices in order to determine the mental health status of a patient and give them feedback in real time. These systems use machine learning algorithms that allow the systems to recognize emotional signals such as variations of tone, word choice and so on, and identify early signs of anxiety, depression or stress. In this case, an AI based app may pick up patterns of negative self-talk and suggest cognitive behavioral therapy (CBT) exercises or suggesting techniques such as mindfulness to help the user with his emotional state. Early detection and intervention of this kind means mental health issues do not escalate and this type can improve long term mental wellbeing (Alexiei Dingli, 2023).

In addition, AI is improving mental health care's accuracy and personalization. AI can use analysis of very large data sets of clinical studies and patients records to discover patterns which connect particular symptoms to the best treatment methods. It permits healthcare providers to plan more focused and efficient treatment in course for the patients. For instance, AI can offer that a patient suffering from treatment-resistant depression could take advantage of a combination of medication and psychotherapy once the same insights are derived from similar cases. These are AI empowered virtual therapists that interact with patients using natural language processing (NLP) in a way that is not too different from human conversation in order to offer emotional support and coping strategies. And these virtual assistants can be programmed to change based on the patient's communication style and emotional state, making it easier and more comfortable and supportive for the patient. Moreover, AI helps monitor mental health with the help of tracking mood changes, sleep cycles, and physical activity, which helps the patient and the mental health professional to garner insights. The ability to monitor mental health in the way, which allows for already continuous change of the mental health care plan in real time ensures more responsive, effective treatment, that allows patients to proactively manage their mental health.

4.5.1 AI for Early Diagnosis of Depression and Anxiety

A new role of AI in depression and anxiety: It is playing an important role in their early diagnosis by analyzing the behavioral, linguistic and physiological data to detect the symptoms of mental health problems before they become bad. Visually assessing these measures of depression and anxiety through traditional methods involves self-reporting and clinical observation, both of which increase the amount of time it takes to identify symptoms and find treatment. On the other hand, things can be constantly monitored by AI driven systems leveraging all sorts of data from patient behavior, social media activity, text-based communication, voice patterns and wearable device data to spot unnatural little lesions of mental health degeneration. By processing large patient profile patient data sets containing information about depression or anxiety, the patterns that are inherent in these large datasets can be learnt by machine learning to predict whether a person is at risk or will have depression or anxiety. For instance, AI can interpret the tone and feeling of someone's word, as well as written text, and can recognize and detect an increase in negativity, hesitation, or social withdrawal, which are indicative of depression. Genetic data is essentially similar data. In a similar manner, AI models can spot indications of anxiety, including increased heartbeat, irregular sleep patterns or changes in activity levels, by monitoring dating coming from fitness watches or smartwatches.

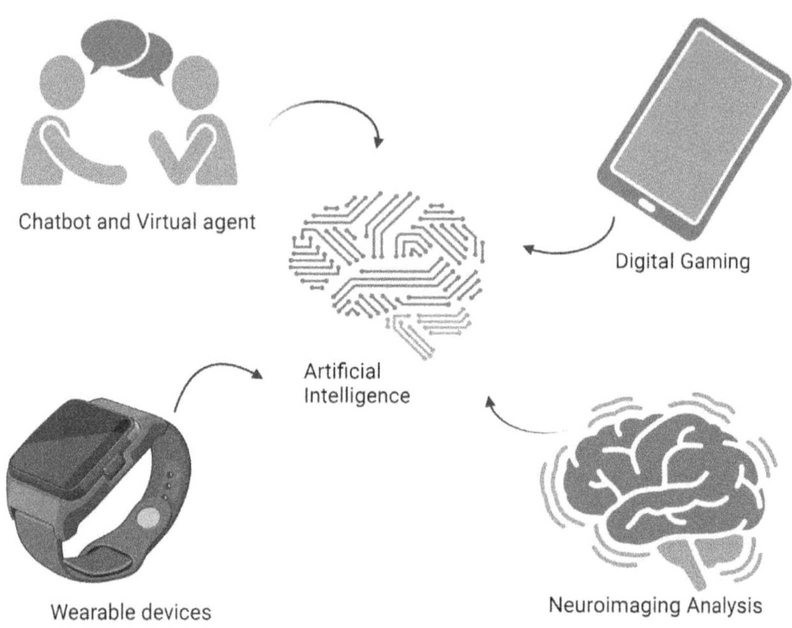

Figure 4.9: Four Pillars of AI in detection of Depression and anxiety.

Source: (Zafar et al., 2024)

Such systems based on artificial intelligence offer a proactive mental health care approach that allows acting on time. If an AI system identifies potential depressed or anxious signs, it can alert the healthcare providers or direct to some self-help resources like relaxation methods, therapy or review of medication. Similarly, AI based mental health apps can communicate with users in an interactive way through conversations that provide emotional support along with coping strategies as per user's specific needs. A chatbot that is programmed with the cognitive behavioral therapy (CBT) principles can, for example, help users go through the structured exercises that challenge the negative thought pattern and help them build emotional resilience. These systems can learn from user interactions, get better at predicting the user's intent, get better at responding, and the process repeats itself. AI monitors continuously and makes an early diagnosis to prevent any mental health problems from escalating, it helps to ease the burden in the healthcare provider and gives better patient outcomes by the timely data-based intervention.

4.5.2 AI in Stress Management and Sleep Optimization

The analysis of physiological and behavioral data about stress, sleep and optimization, and the creation of personal, AI interventions, that can contribute in improving mental and physical health, are transforming AI for stress management. Chronic stress and poor sleep are closely related, with sleep being a major target of chronic stress and stress being a result of poor sleep. By using data from wearable devices, smartwatches or phones, AI driven system can watch over factors like heart rate variations, breathing patterns, sleep cycles or activeness. This is machine learning algorithm that process this data and looks for signs of stress, such as high heart rate, irregular breathing or high levels of cortisol and suggests a particular relaxation technique. Therefore, once AI detects high stress, it can provide ideas such as guided breathing exercises, mindfulness sessions, or even personalized music therapy that coincides with the user's preferences and stress levels. AI based stress management platform also analyses the text and speech patterns of depressed individuals to detect emotional stress and instantly offer support through virtual assistants or therapy prompts. The real time feedback allows the users to know more clearly about their stress triggers and learn to adopt the coping strategies before the stress gets into the escalation stage(Amiri et al., 2023).

Sleep optimization also relies greatly on the role of AI that analyzes the sleep pattern and suggests the best way to optimize sleep to get the best quality. AIS can be used to monitor sleep stages, – light, deep, REM – and detect disturbances due to environmental factors, stress, and even poor sleep hygiene, among other things, as the wearable device we mentioned. AI, therefore, can then suggest making changes in bedtime routines, the temperature or light in a room, and relaxing exercises before sleep. An AI based sleep assistant may pick up that a user is going frequently to bed at 3 a.m., and may suggest cutting down on screen time before or cutting the caffeine that the user consumes. Certain AI systems even go on to control smart home devices in order to make an optimal sleep environment, such as dimming lights or adjusting the thermostat. As it gets deeper into training, the AI models from users' behavior and adapt the recommendations to promote better sleep while reducing stress. Using AI to integrate stress management and sleep optimization and the users can better achieve a balance of the emotional and how they are overall.

Multiple Choice Questions (MCQs)

1. **What is one of the primary benefits of AI-enabled smart medical devices?**

   a. Reducing the cost of medical procedures

   b. Enhancing the visual appeal of medical devices

   c. Real-time data processing and adaptive responses

   d. Increasing the weight of medical devices

2. **How does AI improve emergency response and ambulance coordination?**

   a. By reducing the number of ambulance vehicles

   b. By predicting emergency calls

   c. By providing real-time data analysis and route optimization

   d. By increasing the size of emergency medical teams

3. **In blockchain for secure medical records, how does AI enhance data integrity?**

   a. By increasing data redundancy

   b. By automating the verification of medical data

   c. By reducing patient access to records

   d. By preventing healthcare providers from accessing data

4. **Which of the following is a key advantage of AI in decentralized health data management?**

   a. Centralized data control

   b. Enhanced patient privacy and data security

   c. Increased dependency on a single server

   d. Manual record updates

5. **How does AI contribute to optimizing prosthetic designs?**

    a. By making prosthetics heavier

    b. By using machine learning to customize movement patterns

    c. By increasing the cost of manufacturing

    d. By eliminating the need for patient input

6. **What is one of the major benefits of AI in bioprinting for tissue engineering?**

    a. Reducing the need for healthcare professionals

    b. Enhancing the aesthetic appeal of printed tissue

    c. Ensuring structural accuracy and functional performance of tissues

    d. Replacing traditional surgical procedures entirely

7. **How do AI-driven virtual assistants support mental health and well-being?**

    a. By providing automated emotional responses

    b. By offering personalized support and early intervention

    c. By increasing reliance on human therapists

    d. By replacing traditional therapy methods completely

8. **AI in personalized patient engagement enhances healthcare by:**

    a. Creating standardized treatments for all patients

    b. Reducing the need for doctor-patient interaction

    c. Tailoring healthcare recommendations based on individual data

    d. Increasing waiting time for medical consultations

9. **How does AI help in the early diagnosis of depression and anxiety?**

    a. By analyzing patient heart rate only

    b. By monitoring speech patterns, facial expressions, and behavioral data

    c. By replacing mental health professionals

    d. By increasing patient anxiety levels

10. **What is one of the key contributions of AI in sleep optimization?**

    a. Increasing screen time before sleep

    b. Monitoring and adjusting sleep patterns through data analysis

    c. Reducing user access to sleep data

    d. Recommending increased caffeine consumption

Answers:

1	2	3	4	5	6	7	8	9	10
c	c	b	b	b	c	b	c	b	b

The Future of AI in Healthcare

5.1 Next-Generation AI Models in Medicine

Existing language models have revolutionized how we interact and use powerful AI models for text-based use cases in healthcare. But the practice of modern medicine is chiefly multimodal. Effectively assessing the complete picture of patient health requires moving beyond medical text comprehension to sophisticated AI models skilled of integrating and analyzing diverse data sources across modalities such as medical imaging, genomics, clinical records, and more.

The creation of comprehensive multimodal models has traditionally been hindered by the need for large-scale, integrated datasets and the significant computational power needed to train these models. These barriers have limited the ability of many healthcare organizations to fully leverage AI.

Microsoft Cloud for Healthcare helps to bridge this gap and accelerate AI development. The healthcare AI models, a set of state-of-the-art multimodal medical imaging foundation models, are now accessible in the AI model catalog on Microsoft Azure. They are pleased to announce their debut. These AI models, created in tandem with Microsoft Research and other key partners, are ideal for healthcare organizations looking to prototype, test, and develop AI solutions that meet their unique requirements with minimal investment in computing power and data storage compared to custom-built multimodal models. With healthcare AI models, health professionals have the tools they need to explore the full potential of AI to transform patient care.

Healthcare AI models include:

- **Med-Image Insight**: An embedding model enables sophisticated image analysis, including classification and similarity search in medical imaging. Healthcare organizations and researchers can use the model embeddings and build adapters for their specific tasks, streamlining workflows in radiology, pathology, ophthalmology, dermatology, and other modalities. For example, researchers can explore how the model can be used to build tools to automatically route imaging scans to specialists, or flag potential abnormalities for further review, enabling improved efficiency and patient outcomes.

- **Med-Image Parse**: Designed for precise image segmentation, this model covers various imaging modalities, which include pathology slides, dermatology pictures, ultrasounds, x-rays, CT scans, and MRIs. It can be adjusted for certain uses like tumor segmentation or organ delineation, giving developers a chance to see if AI can be used for more precise cancer detection, diagnosis, and treatment planning.

- **CXR-Report Gen**: Chest x-rays are the most common radiology procedure globally. They're crucial because they help doctors diagnose a wide range of conditions—from lung infections to heart problems. These images are often the first step in detecting health issues that affect millions of people. By incorporating current and prior images, along with key patient information, this multimodal AI model generates detailed, structured reports from chest x-rays, highlighting AI-generated results directly on the images to align with human-in-the-loop workflows. Researchers can test this capability and the potential to accelerate turnaround times while enhancing the diagnostic precision of radiologists. This model has demonstrated exceptional performance on the industry standard MIMIC-CXR benchmark.

These foundational models can accelerate the arrival of groundbreaking AI models that bring intelligent workflows, efficient report generation, and advanced view identification and segmentation to the radiologist experience. In addition to supporting report accuracy, AI can help advance patient care by unlocking new insights from radiology and pathology and genomics, accelerating the discovery of new treatments for disease, and predicting outcomes and optimal treatment plans (Matthew Lungren, MD, 2024).

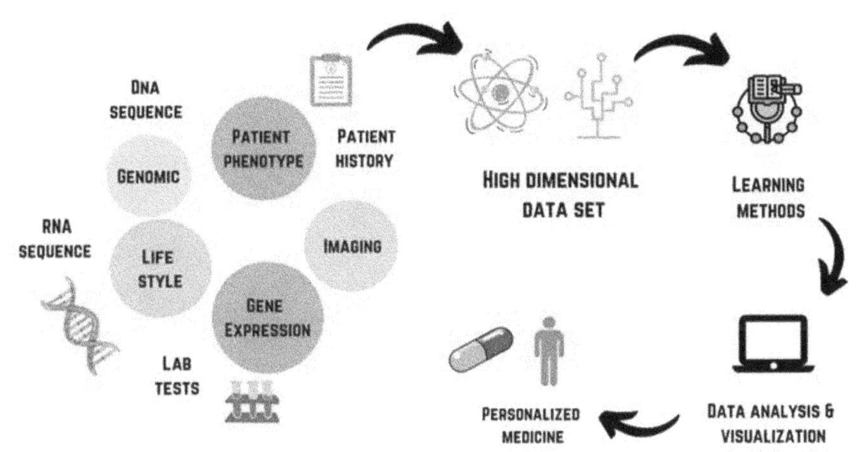

Figure 5.1: Artificial Intelligence for Research in Medicine.

Source: (Peek et al., 2015)

This image illustrates the process of personalized medicine, starting with the collection of diverse patient data including DNA and RNA sequences, phenotype, medical history, lifestyle, gene expression, lab tests, and imaging. This data forms a high-dimensional dataset that is analyzed using machine learning methods. The results are then visualized and interpreted, leading to the development of personalized treatments, represented by a pill and a person, ultimately aiming to improve patient outcomes.

Benefits of AI in medicine

- **Informed patient care:** By incorporating AI into clinician processes, providers can gain crucial context for care decision-making. By providing valuable search results with evidence-based insights regarding treatments and procedures, a trained machine learning system can help physicians cut down on research time—all while the patient is still in the room with them.

- **Error reduction:** There is some proof that AI can make hospitals safer for patients. Improving error detection and medication management are two areas where AI-powered decision support technologies have shown promise, according to a new systematic review of 53 research on the topic.

- **Reducing the costs of care:** The healthcare business stands to save a significant amount of money with the help of artificial intelligence. Opportunities to improve administrative and clinical operations, decrease prescription errors, avoid fraud, and provide personalized virtual health aid are among the most encouraging.

- **Increasing doctor-patient engagement:** Even outside of normal business hours, many patients have enquiries. Even when a doctor's office is closed, AI-powered chatbots can answer patients' most basic questions and direct them to helpful resources. To further assist doctors in identifying health changes that require extra attention, AI has the ability to priorities enquiries and mark information for additional study.

- **Providing contextual relevance:** The ability of AI algorithms to differentiate between various kinds of data based on context is a big benefit of deep learning. For instance, a properly trained AI system can utilize natural language processing to determine which drugs are part of a patient's medical history if a clinical note contains both the patient's existing meds and a new medication that their provider suggests. (Kirilenko, 2020).

5.1.1 Explainable and Interpretable AI in Healthcare

Healthcare costs are headed up all around the world. Contributing to this trend is the fact of increasing life expectancy, soaring rate of chronic disease, and ever-increasing costs for new therapies. This means that it should come as no surprise to scholars who predict a gloomy future for the sustainability of healthcare systems around the world. With AI being able to improve healthcare and make it more cost-effective, the negative impact of these processes should be offset. In practice, however, AIs generally appear as clinical decision support systems (CDSS) to help clinicians reach disease diagnostic and treatment decision choices. While conventional CDSS tends to match the characteristics of individual patients to the existing knowledgebase, AI-based CDSS uses AI models learned from data of patients that have characteristics matching use case in hand. While there's no doubt about AI's potential as a universal solution, it is not. History is replete with things that have progressed along with technology, new questions, and tough stuff. First, certain aspects of these challenges are technology specific to AI, and others that are dependent on the legal, medical, and patient aspects; therefore, there is a need for taking a multidisciplinary approach.

The explain ability of AI is approached from a multidisciplinary perspective in a significant medical AI challenge. An AI-driven system is considered explainable if it enables a human to piece together the reasoning behind a specific AI's predictions. The concept of explain ability is multi-faceted, and, sadly, it's sad that the words used to describe it are not precisely defined. Many people use the words interpretability and transparency interchangeably. As a result, we just make occasional references to explain ability or explainable AI and provide the background readers need to grasp the concept.

Explain ability is a contentious subject with significant ramifications that beyond the technical attributes of AI. Although research demonstrates that AI algorithms can surpass human performance in specific analytical tasks (e.g., pattern identification in imaging), their lack of explain ability has faced criticism in the medical field. Legal and ethical ambiguities about this topic may hinder advancement and obstruct innovative technologies from realizing their promise to enhance patient and public health. However, without meticulous examination of the significance of explain ability in medical AI, these technologies may neglect fundamental ethical and professional standards, overlook regulatory concerns, and inflict substantial harm.

To enhance the dialogue on explainable artificial intelligence in the medical field, this effort intends to highlight the interdisciplinary character of explain ability and its implications for the future of healthcare. Specifically, our research emphasizes the significance of explain ability in CDSS. The uniqueness of our work stems from our examination of explain ability through many perspectives that are often considered independent and distinct. This work has two primary objectives: (1) to deliver a thorough evaluation of the significance of explain ability in CDSS for clinical application; and (2) to conduct an ethical analysis of the implications of explain ability for the integration of AI-driven tools into clinical practice.

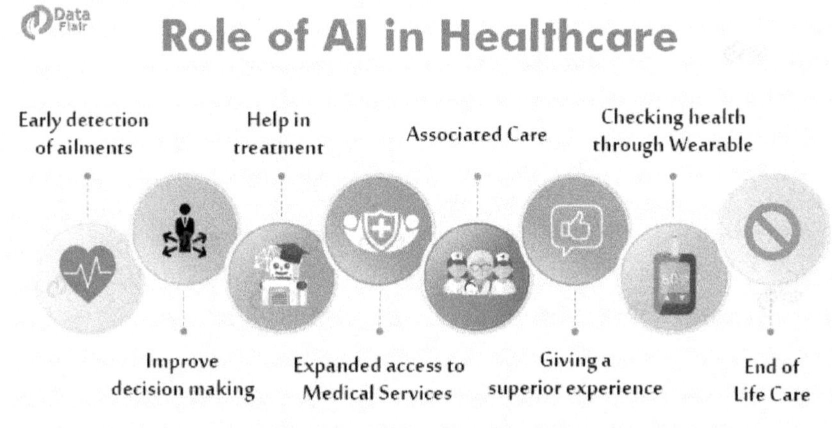

Figure 5.2: Role of AI in Healthcare.

Source: (Secinaro et al., 2021)

This graphic depicts the development of modern healthcare, beginning with the use of wearables for early disease detection, progressing to better decision-making and AI-assisted treatment, increasing access to healthcare, placing an emphasis on interdisciplinary teams providing associated care, providing a better patient experience, and finally, delivering enhanced end-of-life care, all centered around data.

However, because to various sources of inaccuracy, AI systems still cannot deliver 100% accuracy, no matter how much work is put into them. First, it is very difficult, if not impossible, to create a model free of mistakes due to the inherent imperfections of medical datasets (e.g., owing to noise or recording problems). A tee error is an erroneous measurement. Yes, Tus, there will inevitably be some instances of incorrect predictions. Furthermore, AI bias is a major cause of mistake. When AI is biased, it consistently deviates from its predicted prediction behavior, which results in systematic errors. In a perfect world, the training data would accurately reflect the target population of the AI product. Through extensive clinical validation and development on diverse data sources, a primary objective of AI in healthcare product development is to approach this ideal state. Although this makes it easier to minimize AI bias, it will still be very difficult to create AI tools free of prejudice. The presence of bias causes predictions to be inaccurate for patients who do not belong to the training sample. Even with a fully vetted, high-performing AI system, there will be a certain number of faults that doctors and patients may face in the clinical situation. These errors can be either random or systematic.

5.1.2 AI in Quantum Computing for Drug Discovery

For the past five years, cutting-edge tech like quantum computing and artificial intelligence has been booming. Their work has opened up a world of possibilities for innovation, especially in the field of drug development. Companies can take use of the growing computing capacity to run intricate algorithms that can foretell how drugs will interact with one another in the human body. These algorithms can be fine-tuned with the use of AI and machine learning, leading to more accurate predictions of the drugs' actual effectiveness. The potential for these cutting-edge technological advances to speed up the development process, create more realistic simulations of how drugs behave in the human body, and ultimately provide patients with safer and more effective therapies is enormous.

Drug development has historically been costly and time-consuming, frequently requiring decades and costing producers billions of dollars. Although progress has been made, traditional approaches are reaching their limits in addressing the complexity of contemporary diseases. However, quantum computing promises to be able to do computations that are beyond the capabilities of traditional computers.

Personalized medicine could benefit greatly from quantum computing. Researchers are often unable to draw conclusions due to a lack of data from clinical trials. The use of quantum machine learning has the potential to improve the analysis of sparse clinical trial data, enabling researchers to uncover useful insights that could be inferred using classical computers alone. For instance, quantum technologies have the potential to shed light on a patient's individual genetic composition and the potential effects of various treatments.

Quantum computing, AI models, and simulations are used to simulate real-world circumstances using complicated physics-based calculations in medication research. Businesses utilize simulations to tackle real-world problems, and then they optimize solutions in silico using AI and quantum technology. This method saves a lot of time and money by speeding up and simplifying the medication development process. Because of this, we can learn more about materials and molecules, which opens up a world of possibilities when it comes to investigating protein-antigen interactions and more. When compared to more conventional approaches, in silico drug development has many benefits. The first advantage is that pre-clinical asset learning and analysis can be done much more quickly. Researchers can now speed up the process by simply entering real-world problems into a computer, rather than depending only on physical synthesis and testing in wet labs.

The use of AI and generative algorithms has opened up new avenues for drug discovery, allowing researchers to probe protein-antigen interactions, antibodies, and even messenger mRNA. The increased computing capacity and flexibility highlight the possibility of developing superior medications and enhancing patient results.

5.2 AI for Preventive Healthcare and Early Disease Detection

Artificial intelligence has an incredible role to play in preventive healthcare and early disease, across vast numbers of data, advanced machinery learning models, and predictive analytics. AI in healthcare lets healthcare providers predict risks of disease, intervene by taking early action, and enhance patient outcomes by determining patient history, genetic factors, and real-time health information. The proactive approach to this healthcare positively affects the medical systems and increases people's quality of life through timely prevention strategies.

The best application of AI for preventive healthcare has been predictive analytics which involves the processing of huge datasets to predict patterns indicating a possible threat to health. AI models analyze individual's different factors like genetics, lifestyle habits, medical history, and environmental influences to estimate an individual's risk of developing diseases including cardiovascular diseases, diabetes or cancer. For example, AI-based tools can evaluate the levels of cholesterol, blood pressure, etc., combined with trends in physical activity to predict heart disease risks much before the symptoms appear. The information provided explains to healthcare providers which early interventions, like dietary changes, exercise routines and medication plans are recommended to reduce the possibility for severe complications in the future (Dixon et al., 2024).

Improvements in detection accuracy and speed of disease are brought forth thanks to AI in the field of medical imaging and pattern recognition. Due to the fact that traditional diagnosis methods are heavily dependent on human interpretation, misdiagnoses can sometimes occur or it takes too long. These medical images x-rays, CT scan, MRI are populated to the AI driven system Google's DeepMind and IBM Watson Health to analyze them with high precision. But the techniques are trained on huge datasets and fare as well, if not better, at spotting early-stage cancers, lung disease and neurologic disorders as human specialists. For instance, AI can distinguish a breast cancer from mammograms at earlier stages than the traditional methods, which helps to raise the survival rates by early treatment.

With the integration of AI with wearable technology, the continuous health monitoring made possible has been made possible with the use of portable wearable gadgets. There are devices out there, like the Apple Watch, Fitbit and WHOOP, that analyze through the use of AI algorithms heart rate variability, oxygen saturation levels and sleep patterns in order to identify irregularities that may in fact indicate some kind of an underlying health issue. One example is AI powered ECG monitoring in smart watches that can sense atrial fibrillation which increases the risk of stroke and alert the users to seek medical care before a serious event happens. AI driven wearable glucose monitors also help diabetic patients monitor their blood sugar more efficiently and thus mitigate the risk of complications such as the kidney failure or neuropathy.

It is also possible to personalize the risk of some diseases with recent development in AI.AI can go further to analyze the DNA sequences, identify the genetic predisposition to diseases such as Alzheimer's, Parkinson's and a number of types of cancer. Powered by this, preventive strategies against genetic risk can be selected like early screenings for persons at risk or tailored lifestyle changes. Precisely, AI integration in genomics is highly insightful in precision medicine, where individuals' genetic profile can be followed to customize treatments for each patient, for example, improve the effectiveness of interventions conducted.

A NLP powered by AI is playing a key part in monitoring mental health and neurodegenerative diseases at the early stage. Virtual health assistants and chatbots analyze what people say (speech patterns), what they type (text inputs) and this means analyzing what they do (behavioral cues) to detect possible signs of depression, anxiety or schizophrenia. For instance, AI algorithms can recognize linguistic markers in a patient's speech indicating cognitive decline so that the condition including Alzheimer's disease can be tackled early. Like this, mental health apps also use AI driven conversations to give emotional support through real time conversations to users to manage stress and anxiety before it becomes severe issues.

By understanding how various compounds act on diseases, AI is helping drug discovery and the development of preventative treatments to run faster as well. But an example of capability this was particularly clear for during the COVID 19 pandemic where AI driven models allowed researchers to identify potential drug candidates and vaccine formulations at a pace which has never been known by them. The development of new preventative treatments for age related diseases, as well as vaccines against infectious diseases, are powered by AI as part of the same era's suite of applications. AI shortens the research time and the research expenses by streamlining the process of research into the development of new medical solutions, which in turn makes preventive medicine available and effective.

5.2.1 AI in Cancer Screening and Early Diagnosis

The domains of early cancer diagnosis and AI are undergoing fast development and experiencing significant intersections. The United Kingdom's national registry data reveals a strong correlation between cancer stage and 1-year cancer mortality. Moreover, for some subtypes, there is an incremental reduction in result for each stage that increases. In the case of lung cancer, for instance, 5-year survival rates after stage I disease resection are 70-90%; nevertheless, the overall rates are currently 19% for women and 13.8% for men. In 2018, 44.3% of English cancer patients were diagnosed with an early stage (I or II) disease; however, for gastric, pancreatic, esophageal, and oropharyngeal cancers, the proportion was less than 30%. The long-term strategy of the National Health Service (NHS) stated a national aim to increase the percentage of early diagnosis to 75% by 2028. Several international bodies, such as the World Health Organization (WHO) and the International Alliance for Cancer Early Detection (ACED), have acknowledged early diagnosis as a critical priority.

While many studies highlight the efficacy of screening in early cancer detection and mortality, there is high degree of discordance of patient selection and risk–benefit tradeoff and concerns over the former being 'one size fits all' especially when it doesn't follow personalized medicine aims. There are key challenges of patient selection and risk stratification with screening programmers. In the near future, AI algorithms to process huge quantities of multi-modal data to highlight previously difficult to identify signals could have a role in enhancing this process. Additionally, AI can directly drive cancer diagnosis as it can prompt investigation or referral of investigated screened individuals according to clinical parameters and automate clinical workflows where capacity is constrained. In this review, we discuss the possible the ways that AI might be used for early cancer diagnosis in symptomatic and asymptomatic patients to examine what types of data can and cannot be used, and the likelihood of benefits occurring in those areas within the next few years.

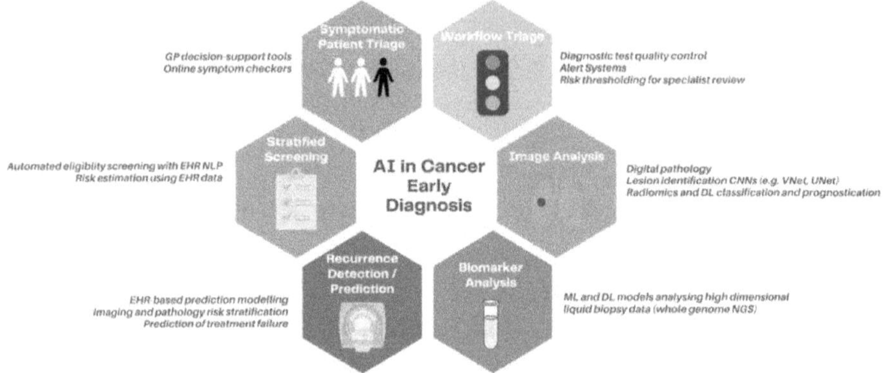

Figure 5.3: AI in Cancer Early Diagnosis.

Source: (Hunter et al., 2022)

Medical uses of artificial intelligence for the detection of cancer at an early stage. Common Synonyms: Natural language processing (NLP), "general practitioner," A patient's medical history stored digitally NGS stands for "next-generation sequencing," ML for "machine learning," and DL for "deep learning."

Opportunities in AI-Driven Cancer Diagnosis

1. **Early Detection and Improved Screening:** Early detection of cancer at an earlier stage can be achieved by AI, which can hear that and save a lot of lives. Medical imaging data such as mammograms, CT scans, MRIs and histopathology slides can be analyzed by 'deep learning' algorithms and the abnormalities in the data identified which may be overlooked by human radiologists. Said to be promising are other AI based screening tools for breast, lung and cervical cancer in clinical trials. To illustrate, Google's DeepMind AI model has successfully outperformed radiologists in the detection of breast cancer but at a higher accuracy than they can achieve, in terms of reducing false positives and negatives.

2. **Enhanced Diagnostic Accuracy and Reduced Human Error:** Quick and precision diagnostic analysis of large amount of patient data is a capability of the AI systems that minimizes the chances of diagnostic errors. Currently, human evaluation is utilized in traditional pathology, and interpretations are prone to variability. Convolutional neural networks (CNNs), along with other AI powered tools have been trained to classify benign and malignant tumors with the use of large datasets that have been annotated. For example, IBM Watson Health's oncology platform can assist oncologists in most accurately making a diagnosis by cross referencing patient data with global medical literature.

3. **AI-Powered Personalized Cancer Treatment:** Precision oncology is enabled by AI from a genetic, molecular and histological data to tailor treatment to individual patient. Using genomic sequencing, AI makes it possible to pinpoint mutations, as well as biomarkers, that make a patient prone to a certain treatment. Tempus and Foundation Medicine built AI models that enable matches between cancer patients, and the most effective targeted therapies, based on genetics. The personalization enhances the effectiveness of treatment with less toxicity.

4. **AI for Histopathology and Digital Pathology:** By combining digital pathology with AI, samples are being analyzed differently. Patterns indicating cancerous changes can be identified in gigapixel resolution histopathology slides using AIs that process the slides. Historical pathologists rely on AI powered histopathology tools to guide the highlighted areas of concern, speed up diagnosis and increase reproducibility. Their own studies confirmed that AI models—such as those the company developed—can outperform humans at spotting cancerous tissue in biopsy specimens.

Challenges in Implementing AI for Cancer Diagnosis

1. **Data Availability, Quality, and Bias:** To reach reliable performance AI models are highly reliant on a lot of high quality, annotated medical data. The problem is that we do not

have access to various and sufficient datasets. Most of the AI models are trained on the restricted, region-specific datasets which could inherently have the potential of the biases and less or no generalizability. For example, suppose we train an AI model on African patient data in particular, and found it would not accurately diagnose cancer on African patients due to an underrepresentation of people from that ethnic group in the image data set. To address this, some efforts to mitigate this include the global data sharing initiatives and federated learning.

2. **Lack of Explain ability and Trust in AI Decisions:** Some point to "black box" problem of devising ways to explain to physicians how the AI came to their decisions as one of the larger hurdles blocking adoption of AI in oncology. However, there is still reluctance on the part of the clinicians to trust the AI based diagnoses without knowing the reasoning behind them. XAI models are necessary to build trust with healthcare professionals, that is, the ones that are developing the AI systems. Efforts are being made towards integration of attention maps and visualization techniques for bigger interpretability.

3. **Integration into Clinical Workflows:** There is a logistical issue to the seamless incorporation of AI into existing healthcare infrastructure. Legacy systems that are not made with the AI tools in mind are widespread in many hospitals and diagnostic centers. Implementation of AI demands substantial investment in infrastructure, compatibility of software, as well as the training of their staff. Real world validation and significant clinical trials are also needed before using AI tools are widely practiced in oncology departments.

5.2.2 AI-Driven Predictive Health Analytics

The healthcare industry has a rich tradition of anticipating future needs. Many have pondered the question of how predictive analysis might improve healthcare in light of current developments in AI. To enhance healthcare services in places with a high risk of disease, big data analytics, artificial intelligence, and machine learning models can be invaluable. For instance, data scientists can benefit from a deeper understanding of the many aspects impacting human health by combining AI with predictive analytics.

To illustrate how AI might enhance population healthcare management, let's look at a few examples:

1. **Predicting chronic and infectious diseases:** Healthcare professionals can utilize predictive analytics to make highly probable sickness predictions based on a range of data points, such as the patient's birthplace, place of employment, lifestyle choices, and local environmental factors. As a result, doctors will be better able to assess their patients' vulnerability to chronic diseases and devise strategies to minimize their risk. A number of chronic diseases, including diabetes, heart failure, and COPD, are experiencing declining rates, which is a positive sign for the sector.

2. **Streamlining patient throughput and workflows:** Predictive analytics driven by AI can help with resource allocation by predicting changes in patient flow. These projections can optimize staffing levels to improve patient care, extrapolate demands from patient data to guarantee appropriate bed allocations, and improve healthcare facility employment rates.

3. **Predicting hospital readmissions:** The use of big data analytics, in conjunction with powerful computing resources and machine learning, allows for the prediction of catastrophic epidemic situations. Weather reports, reported cases, population density, economic profile, and other data sources can potentially be used to anticipate the spread of contagious diseases.

4. **Analyzing patient deterioration:** Data scientists can use predictive factor analysis to learn about and track the development of diseases, as well as to foretell potential negative drug or therapy side effects, allowing them to intervene with patients at the optimal moment.

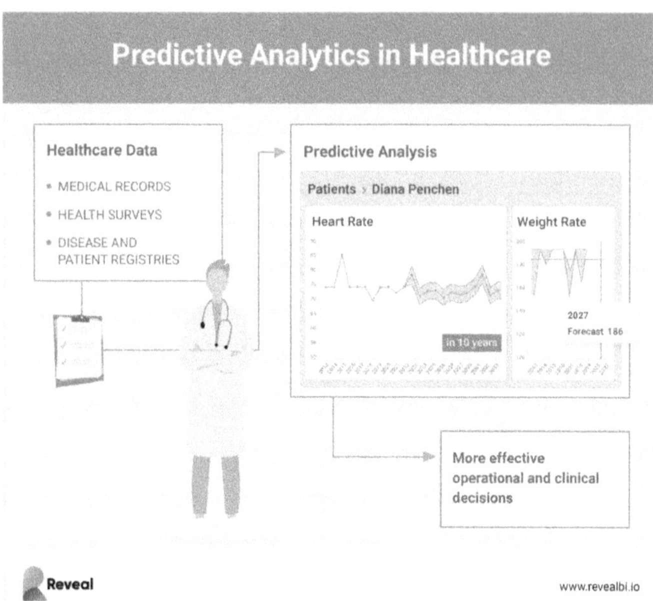

Figure 5.4: Predictive Analytics in Healthcare.

Source: (Van Calster et al., 2019)

Nowadays, there are a lot of different places to find health records and other related information. Some of these places include physical exams, lab imaging, electronic records from medical equipment, and even smart devices. Healthcare practitioners and researchers have access to limitless resources for better diagnoses and treatments because to AI's continual assessment and analysis of this data.

The field of machine learning has reached maturity. The combination with AI makes it capable of delivering strong, practical insights. With the use of these new technologies and contemporary data sets, population health management can reach its end objective: improved health outcomes.

5.3 AI's Role in Global Healthcare Accessibility and Affordability

The current business activities (e.g., providing quality healthcare, increasing accessibility and affordability, and cutting costs) are now being transformed by Artificial Intelligence (AI), addressing the key challenges. But healthcare systems all over the world, especially in low- and middle-income countries are not able to provide sufficient medical infrastructure and trained professionals while high treatment cost. AI makes healthcare more equitable and cost effective by automating tasks, improving diagnostics, remote patient monitoring, among other things.

With scalable and efficient medical solution, AI can actually bridge the gaps in underserved regions in health care. With the help of AI, virtual consultations through telemedicine platforms create an environment where patients no longer have to visit healthcare physicians for consultations where they reside are far away. Babylon Health and Ada Health are two such AI driven chatbots and virtual assistants that offer preliminary diagnostics through analyzing symptoms and offering medical advice. These technologies decrease the burden on healthcare facilities without diminishing the time patients wait for guidance.

Furthermore, AI driven mobile clinic equipped with diagnostic tools are being used in rural and remote areas to deliver a medical service to population with poor access to healthcare. These clinics are using AI for automated screenings, disease detection, and digital documentation and do not rely heavily on the human healthcare providers. What comes to mind when considering a more industrial type of application for AI in the healthcare space is when AI powered telemedicine is bringing the millions of people in countries like India and sub-Saharan Africa medical treatment at a time when there are shortages of doctors and specialists.

Lowering healthcare costs is important and AI has done this through improved operational efficiency, minimizing of unnecessary medical procedures, and reducing diagnostic errors. Predictive analytics is one of the primary ways AI can accomplish this for hospitals, providing predictions of future patient admissions, better routing of resources, and prevention of disease contamination. AI analyses large datasets in hospitals to help hospitals prepare for patient influxes, thus addressing the shortage of staff and medical supplies better.

One more crucial cost saving involves the automation of administrative tasks. The help of AI-powered tools allows medical documentation, appointment scheduling, and billing processes go faster and are less cumbersome for healthcare staff. The implementation of AI-driven automation in hospitals has improved efficiency thereby reducing internal administrative expenses, and funds are realigned into patient care. Apollo Hospital in India has used AI to reduce the workload of staff, manage patients better, and are better in treatment outcomes. Further, the hospital plans to expand its AI investment to increase cost benefits of its AI adoption.

With the help of AI, diagnosis is significantly diluted and has become less expensive and time-consuming. Giant medical datasets can be mined by machine learning models to indicate with high accuracy which diseases include cancer, tuberculosis, and diabetic retinopathy. One example is when Google's DeepMind AI has already beaten human radiologists to detect breast cancer from a mammogram and it has resulted in early detection and reduced treatment cost. Medical imaging diagnoses are made more accurately and reliably but also with less need for expensive follow-up tests.

AI also is very important in personalized medicine, tailoring treatments to a patient at the genetic, environmental, and lifestyle levels. Patient data is analyzed by AI algorithms recommending best treatment plans for patients which would otherwise require reducing the use of trial-and-error approaches that can be rather expensive, both for patients and medical organizations. AI improves treatment accuracy and decreases the chances of adverse drug reactions, which enhances patients' lives and cuts long-term healthcare costs.

The cost and time to develop new treatments with AI have been reduced in drug discovery. Typically, it takes over a decade and billions of dollars to develop a drug with traditional drug development; however, AI-supported simulations and predictive modelling processes have identified potential drug candidates faster. In the times of the COVID-19 pandemic, AI played a big role in speeding up vaccine and antiviral drug research, showing possibilities of much cheaper and more promising healthcare to approach new diseases. By addressing the low medication cost problem, it states that AI-driven pharmaceutical development has made essential medications available to patients around the world.

Similarly, AI is also helping to prevent the risk of chronic diseases and suggests lifestyle changes. For example, a wearable device such as a smartwatch gets to use AI algorithms that can track heart rate, blood oxygen level, and physical activity, and give real-time insights into an individual's health. AI-based preventive healthcare helps prevent costly hospitalization by allowing early intervention for the disease of diabetes, high blood pressure and cardiovascular diseases.

5.3.1 AI in Bridging the Gap in Rural and Underserved Areas

Quality healthcare in rural and underserved areas is restricted by a lack of medical infrastructure and also the shortage of healthcare professionals and geographical barriers. Just as AI promises to become an important forerunner in solving such challenges, it also provides fresh ways to narrow healthcare gaps and to assure that remote populations get fairly a hit of timeliness and a more efficient form of medical care.

The incorporation of AI in telemedicine has tremendously improved it, thus helping to access healthcare services in rural areas remotely. AI-based platforms enable virtual consultation with healthcare providers, they allow the healthcare providers to diagnose and treat a patient without actually having to be present. This dovetails not only by reducing travel time and costs for patients but also by releasing the health care facilities from the overstretched facilities burden. For example, the role played by some AI-driven telemedicine platforms in facilitating ICU-level care supply from remote healthcare centers has facilitated their reach to critical care services.

The timely diagnosis is essential for proper treatment, however, rural distance away from the diagnostic infrastructure is a limitation. This gap is addressed by AI because it analyzes medical data in great accuracy. Digital medical imaging such as radiology and pathology scans can be correctly and faster interpreted by AI powered tools than the traditional methods. Some studies have demonstrated that adding automation based on AI image analysis coupled with expert reading leads to improved performance.

Rural areas are being revolutionized in terms of how health conditions are managed by wearable health monitoring devices that are powered by AI. Such devices with AI can be able to monitor vital parameters like heart rate and blood oxygen levels and send real-time alerts to doctors regarding health alerts. Since it is continuous monitoring, hospitals are reduced and early intervention can be done. For instance, thinking about rural patients, AI-powered solutions facilitate mobile app access to important health conditions and how they can be managed information.

A shortage of well-trained medical professionals is often seen in rural areas. Remote medical training is made possible through AI-powered learning platforms and virtual simulations to keep healthcare workers in regions equipped with the latest medical knowledge and strategies. As well, AI-driven clinical decision support systems help rural healthcare providers by providing real-time recommendations for diagnosis and treatment resulting in a reduction of misdiagnosis.

Logistically, it is a task to ensure the availability of essential medical supplies in rural areas. IT is rolling into healthcare, making an inventory of inventory levels as well as demand prediction. AI has supported in optimization of healthcare logistics, such as medical goods and apparatus administration, and remote patient observing to upgrade operational proficiency and expand patient result.

In rural areas, there is a stigma associated with mental health services which makes it unavailable to those around them. A virtual counseling platform that benefits from AI chatbots is becoming more available. By these platforms, people from remote areas can have confidential mental health care without visiting the in-person, to meet the rapidly spreading burden of mental disorders in underserved groups.

However, while this has great potential, there are significant challenges in applying AI in rural healthcare. The most important problem is the digital infrastructure and lack of internet connectivity in most of the remote areas, which is the base for AI-powered healthcare solutions. Virtually, there are additional concerns concerning data privacy and security to guarantee patient confidentiality. Additionally,

AI systems should be developed to address health disparity to improve care access equity so that AI models suggest accurate and fair healthcare recommendations for all populations.

Continuing advancements in AI in rural healthcare indicate that there is a promising future of AI in rural healthcare with more possibilities to enable more accessible and quality healthcare. AI-driven predictive analytics can help healthcare providers predict health problems before they transition into serious situations and suggest more proactive care strategies. Healthcare systems can successfully close the healthcare gap in rural and underserved areas by employing AI responsibly to make sure that all people, wherever they are, have access to high quality medical care.

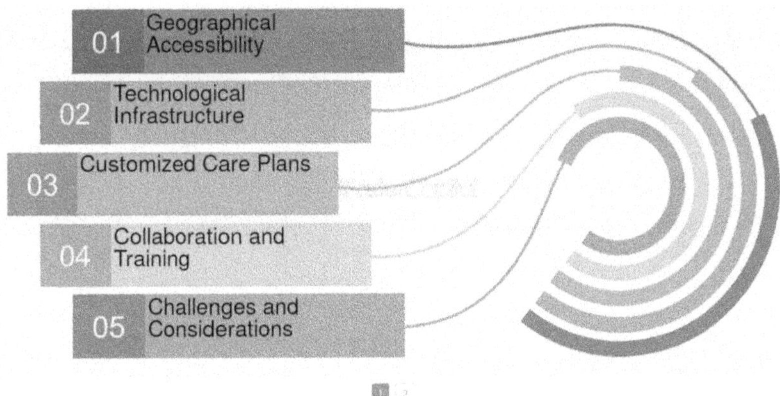

Figure 5.5: Bridging the Gap in Rural Areas.

Source: (Batool & Lopez, 2023)

1. **Geographical Accessibility:** AI enhances geographical accessibility by making it possible to do remote consultations and telemedicine, that means patients in remote or underserved area can see a doctor without having to travel by miles. The analysis of the gaps in the healthcare sector is done by AI and suggests targeted interventions to address resource distribution and patient care.

2. **Technological Infrastructure:** AI integration in healthcare is enabled by strong technological infrastructure which includes secure data storage, real time data access as well as integration

with hospital management systems. Large data volumes and speed in making decisions are handled due to cloud-based platforms along with advanced computing power.

3. **Customized Care Plans:** AI can extract medical information, genetic data, lifestyle factors from medical records, and use this information to derive a personalized care plan. This enables predictions of treatment responses, detachment of care, and improvement of recovery outcomes. Tailored rehabilitation and follow up care can be done by continuous monitoring.

4. **Collaboration and Training:** In healthcare AI there is need for collaboration of medical professionals, developers of AI and data scientists. To effectively adopt and use AI tools, offer training on these tools. Such communication and sharing of knowledge greatly contribute to the development of practical AI solutions that are clinically appropriate.

5. **Challenges and Considerations:** One challenge to AI adoption is that data privacy, algorithmic bias, as well as compliance with regulations are things that tie the arms of AI adopters. Training is needed to get healthcare professionals past the resistance to AI. Important for smooth implementation are also the cost management as well as the access to AI solutions on equitable terms.

There is typically a dearth of healthcare practitioners, fewer healthcare facilities, and broken healthcare infrastructure in rural areas. The populations have been disproportionately impacted by inadequate healthcare facilities and uneven healthcare delivery for many years, leading to a worsening healthcare imbalance. Due to factors such as geographical remoteness, transportation constraints, lack of healthcare awareness, inadequate health workforce, and financial limits, they encounter barriers to healthcare access. However, healthcare providers in rural areas may better grasp the healthcare gap and train themselves to use AI effectively to close it. This will pave the way for more widespread adoption of technology in these areas. The use of AI has the potential to overcome these challenges and increase the availability of healthcare services in rural areas.

In healthcare, AI is more of a supplement that improves results than a substitute for human doctors and nurses. Nevertheless, there are ways to guarantee sufficient healthcare service through the use of these technologies. With the use of AI, telemedicine has become more accessible, opening up healthcare to people living in remote locations. Furthermore, diagnostic imaging and screening have been bolstered by AI tools, allowing healthcare providers in remote areas to better diagnose and manage health issues. Additionally, AI has helped optimize healthcare logistics, which includes managing medical equipment and supplies and remote patient monitoring. These advancements hold great potential for better patient outcomes, more operational efficiency, and lower costs. Numerous examples show how AI tools have been successfully integrated into healthcare in rural areas. However, AI could revolutionize healthcare delivery in rural places within the next decade by making services more accessible, affordable, and of higher quality.

5.3.2 *AI-Powered Affordable Healthcare Solutions*

AI in the healthcare space is transforming the industry in terms of how one can access, afford, and undertake medical services, to make them more accessible. AI-based healthcare solutions for use are powered by highly advanced algorithms, machine learning, and data analytics aimed at improving diagnosis, treatment, patient monitoring, and efficiency of operations making quality healthcare more affordable and accessible to all, and more so, to the people in the completely underserved regions.

Given the complexities that global healthcare systems are coping with including rising costs, aging population, shortage of medical professionals, and the need to enhance efficiency as well as safe and effective care delivery resulting in better outcomes, AI presents an evocative solution. AI is changing the way healthcare is delivered and predictive analytics for disease prevention, diagnostics are done with the help of AI, robotic-assisted surgeries, and virtual health assistants among other things. Not only are these innovations improving patient outcomes, they are also costing less and therefore are more affordable because of them, essential services become more affordable.

The AI-powered solution helps in decreasing the cost by optimizing administrative tasks, lowering diagnostic errors, and tailoring treatment plans according to the data related to the patient. NLP technologies allow for automation of documentation, reducing the burden on the healthcare professional, and allowing him to focus on patient care. As it also, AI-based Telemedicine Platforms do remote consultations as they do not need expensive hospital visits especially in rural and remote areas.

Besides, AI facilitates the basic pharmaceutical research and drug development while accelerating discovery of inexpensive medications and optimizing supply chain management for essential drugs. Wearable devices powered by AI can be used to encourage individuals to monitor their health in real-time, and people can practice early diagnoses and preventative care.

However, while there are many benefits that AI brings to healthcare, adopting AI in healthcare brings along some challenges such as ethical issues, privacy issues, regulatory issues, etc. and also the need for professionals who can interpret the information that AI is generating. Only with durable partnership among the policymakers, healthcare providers, technology companies and regulatory authorities can these challenges be addressed and our AI-powered and ongoing improvements can be rolled out confidently.

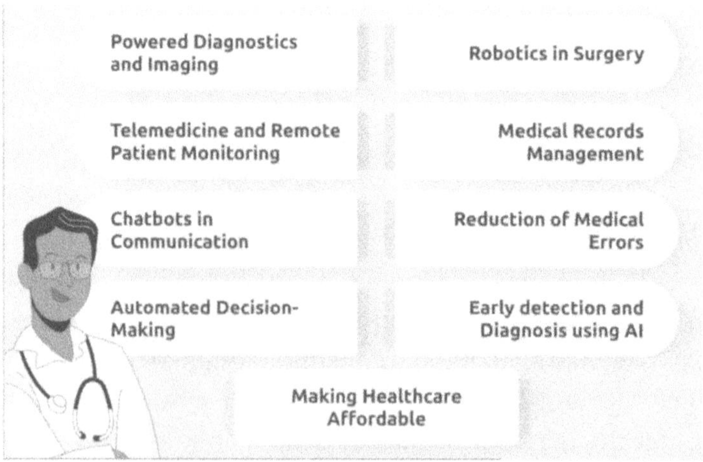

Figure 5.6: Solutions of AI in Healthcare.

Source: (Väänänen et al., 2021)

The figure illustrates multifaceted impact of technology on healthcare, with "Making Healthcare Affordable" as the central goal, achieved through various applications like "Powered Diagnostics and Imaging," "Telemedicine and Remote Patient Monitoring," "Chatbots in Communication," and "Automated Decision-Making," alongside advancements in "Robotics in Surgery," "Medical Records Management," "Reduction of Medical Errors," and "Early detection and Diagnosis using AI," all visually supported by a welcoming depiction of a healthcare professional.

The Need for Affordable Healthcare Solutions

Although healthcare affordability has been addressed in recent times, global healthcare is still a big challenge and in smaller countries where affordability is an issue, the accessibility to quality care is another challenge. Aside from, it is a burden to patients and to healthcare systems in developed nations, the rising costs of medical treatments, hospital stays, and pharmaceutical drugs. These high costs are due to:

- **Expensive Diagnostics and Treatment:** Traditional medical diagnostics, viz. MRI scans, genetic testing, and especially specialized treatments, are very costly and thus inaccessible to the low-income population.

- **Workforce Shortages:** This leads to increased patient load, long waiting times and high fees of consultation due to the shortage of healthcare professionals, such as doctors, nurses and specialists.

- **Operational Inefficiencies:** The administrative bottleneck, redundant paperwork and inefficient use of resources bring the total cost of healthcare service.

- **High Pharmaceutical Costs:** This is a costly, time-consuming process and produces expensive medications which not all patients can afford.

- **Lack of Preventive Care:** Most healthcare systems pay more attention to treating diseases instead of preventing diseases and its cost is very high for the diseases which are preventable through healthcare.

Often relying on artificial intelligence, AI-powered healthcare solutions address these issues through automation of processes, increase in diagnostic accuracy, remote consultations, and better treatment plans thus enabling to make healthcare more cost-effective.

AI-based diagnostic tools are extremely fast and affordable in diagnostic costs. ML algorithms are used to analyses medical images (X-rays, etc.) with great precision to detect early the conditions of cancer, tuberculosis, neurological disorders, etc. The diagnostic platforms powered by AI such as Google's DeepMind and IBM Watson Health have shown to verify their accuracy far more than human specialists can. AI minimizes diagnostic errors and makes fewer expensive tests needed since more tests do not mean less accuracy, it means more costs, and AI is much cheaper.

However, telemedicine has become an effective option to enable the provision of healthcare to rural and underserved areas. Some initial consultations are offered by Virtual assistants and chatbots running on AI, such as Babylon Health or Ada Health, which analyze the symptoms and direct patients to the proper care. AI-integrated wearable devices allow for the usage of remote patient monitoring in which patients can remotely monitor chronic diseases such as diabetes and hypertension, reducing the need for frequent hospital visits. This cuts time and money costs as well as lightens the load of overburdened healthcare facilities.

5.4 AI and Human Collaboration: The Future of Medical Practice

AI and human expertise are reshaping the future of medical practice. Healthcare dictated by AI-driven technologies is advancing at an impressive speed and promises to provide new capabilities in establishing diagnosis, predicting, being robotic for surgeries and implementing customized treatment plans. Yet, although AI is becoming more and more important, humans are still irreplaceable when it comes to doctors, nurses, and healthcare professionals thanks to their thinking, ethically sound decision-making, empathy, and patient cantered approach. Instead of replacing healthcare professionals, AI is working collaboratively with them to provide a better quality of care (Chakraborty et al., 2024)robust, and efficient machine learning (ML.

At this point, medical practitioners and AI are coexisting in a syntonic where technology becomes a crutch and addition for human judgment and not a replacement. For example, AI algorithms capable of analyzing hundreds of millions of logs of medical data can identify patterns and help in diagnosing diseases with huge accuracy. But human expertise is necessary to understand the meaning of an AI's insights, the right context for a diagnosis and how to make ethical treatment decisions. As complex cases tend to be, a customized, nuanced judgment, and emotional intelligence combination is critical, and this collaboration is extremely valuable in that regard.

Better efficiency is one of the key benefits of the potential of AI-human collaboration in medicine. AI administrative tools decrease the amount of paperwork required for doctors and nurse and they have more time to take care of patients. Virtual assistants and chat bots powered by AI are used to field patient inquiring and triage cases, and to schedule for appointments. Moreover, AI helps in accelerating drug discovery in medical research, analyzing genetic data for a 'precision medicine', and detecting various disease risk factors.

Nevertheless, there are still many unanswered questions concerning the ethical application of AI, addressing algorithmic biases, safeguarding data privacy, and preserving trust between healthcare providers and patients, despite the immense potential of AI. To successfully integrate AI in medical practice, clear guidelines, permanent medical education and policies promoting the human control are needed.

Ultimately, the most influential encounters in the future medical practice will be those between AI and human expertise. Without making the systems that humans use inherently moral, or trying to transmit moral intent, technologists need to be ready to repair systems that do unfair or unjust things. This is where AI and human doctors together can revolutionize healthcare and: personalize it, proactively, and, above all else, to the patient.

Key Drivers of Change in Medical Practice

Following, are several factors which could still determine the future of medical practice.

- **Advancements in AI and Machine Learning:** Deep learning algorithms can analyze medical image, make accurate prediction of disease progression and prescribe personalized treatment to a given patient. AI based diagnostic platforms hold out a superior performance in detecting such conditions as cancer and cardiac diseases.

- **Telemedicine and Remote Healthcare:** However, the COVID 19 pandemic has fast tracked the use of telemedicine, which has made patients consult doctors virtually. Today, there are AI based chatbots and virtual assistants that perform initial medical assessments to decongest the workload of health care professionals, with accessibility.

- **Precision Medicine and Personalized Treatment:** Genetics, genomic and big data analytics in AI assists in making customized plans on the basis of patient genetic disposition along with the lifestyle and health history of the patient. This method is also used to improve treatment efficacy and lowering the number of unnecessary procedures and cost.

- **Robotic-Assisted Surgeries and Automation:** Precision procedure is made better with the aid of AI dictated surgical robots which are less likely to cause complications in complex surgeries. AI helps by giving real time data, improving accuracy and reducing patient recovery time and while human surgeons are still in control.

- **Healthcare Data and Predictive Analytics:** Predictive analytics fueled by AI serve as the gift that helps prevent its crisis: they enable predictive interventions in advance of any health symptoms appearing. AI is used to optimize resource allocation, predict patient admission rates, and better support organization's muscle of culture.

The Role of Human Expertise in AI-Driven Medical Practice

However, despite the capabilities of AI, humans in the professional medical practice cannot be replaced due to following reasons:

- **Ethical Decision-Making:** The moral and ethical reasoning necessary to make life and death decisions such as end of life care and emergency prioritization, is not present in AI.

- **Empathy and Communication:** Human interactions, emotional support and reassurance are what patients' value and this AI cannot really provide.

- **Clinical Judgment:** Although AI can make suggestions of treatments given data, doctors would still interpret AI generated insights with respect to the patient's overall health status.

- **Handling Uncertainty:** Currently, humans are needed to interpret AI models with data that would otherwise be used in other industries rare diseases, unexpected complications, ambiguous symptoms or even just not enough others

5.4.1 Augmenting Medical Professionals with AI Assistance

The healthcare industry is being revolutionized due to fast progressing artificial intelligence (AI) is not that it is replacing medical professionals, but instead, it is augmenting their capabilities. Using artificial intelligence, those tools are helping doctors, nurses, and other medical personnel provide more efficient, accurate and effective healthcare services through helping doctors, nurses, and other healthcare professionals with powerful analytical, diagnostic, and operational support. Where healthcare is delivered is being changed by AI, such as in assisting medical imaging and predictive analytics to streamline administrate processes, have positive effects on patient outcomes, and reduce the load of medical practitioners (Bajwa et al., 2021).

They live in today's fast medical world where professionals are expected to meet higher demands for patient loads, complex disease patterns, and administrative insists on their time. One of the solutions is AI-powered assistance that automates repetitive tasks, analyses vast amounts of medical data, and offers evidence-based recommendations that can help healthcare providers spend more time caring for patients. All of this is being ML algorithms, "natural language processing" (NLP), and "robotic process automation" (RPA) are now being combined into the various aspects of medical practice of using EHR management to AI-aided surgeries.

By combining AI with human expertise, medical professionals gain more informed decisions, without that human touch when interacting with patients is lost. Wearable devices and real-time analytics can also be in place to use AI in monitoring patient progress and optimizing treatment plans, as well as in recognizing at earlier stages the diseases. In emergency medicine, AI-based solution helps in physicians' decisions in prioritization of critical cases, shortening survival time with faster response time.

There is a severe scarcity of healthcare workers, which is having an effect on patient care, operational efficiency, and the quality of services provided. Healthcare organizations are facing growing challenges in meeting patient needs. One innovative solution is staff augmentation, which may help fill in skill gaps. AI is helping medical personnel by optimizing diagnostic accuracy, automating routine procedures, and facilitating real-time decision-making. This way, patients still get high-quality care without medical mistakes.

Healthcare organizations may optimize productivity, reduce operational costs, and enhance patient outcomes by integrating AI into workflows. This is all done without sacrificing the crucial human element in medical treatment. The need for AI-powered solutions to keep healthcare workers employed is growing. Automating administrative activities and aiding with complex medical diagnoses are just two examples of how AI staff augmentation is changing the face of healthcare, fostering innovation, and improving service delivery.

- Investing in the AI staff augmentation in the field of healthcare, not only solves the problems of the lack of workforce, but also improves diagnostic accuracy, reduces costs, improves patient experience, meets compliance, and can scale as needs grow.

- A subset of healthcare staff augmentation roles for AI includes virtual health assistant AI, AI-enhanced medical transcription/documentation AI, AI-enhanced radiology diagnostics, predictive analytics, AI drug discovery research, and remote patient monitoring.

- Key challenges involved in AI staff augmentation in healthcare include data privacy/security, integration of legacy system, AI bias/accuracy, resistance to adoption, regulatory/ethical concern, and scalability/maintenance cost but these are manageable by having robust security measures, IT staff augmentation, bias reducing strategy, education/train, ethical framework, and cost-effective scaling.

- Considering AI staff augmentation in healthcare includes determining gaps in workplace, choosing appropriate AI technologies; seamless integration into the existing systems; AI staff training for new used equipment; ensuring compliance and security of collected data; and appointing AI experts for creating custom solutions.

5.4.2 AI in Medical Education and Training

Artificial Intelligence is increasingly now integrated into medical education and training so that future healthcare professionals learn to practice and acquire clinical expertise. AI-assisted tools augment old-age teaching tools and enable personalized learning through provision of personalization by imitating the body system structure, real-time feedback, and immersive simulation. This leads to advancements that enhance the efficiency and effectiveness of medical education so that healthcare professionals reach disparities in diseases and are efficient in performing procedures, and making critical decisions of patient care (Elendu et al., 2024).

AI applications like virtual simulations instructive tutoring systems and AI-based diagnostic mechanisms that allow students to train scenarios in a safe environment are being used in medical training. The machine learning algorithms calculate the student's performance, know the student's areas of lack in knowledge and then give educational content that is related to the student's requirements. It personalizes training for the learners, the way it trains them is right according to their competency and decision-making speed.

Among the most important inventions in medical education is the usage of VR (virtual reality) and AR (augmented reality) simulations assisted by AI. These technologies have also meant that no one lives or dies; medical students and residents can practice or train in surgeries without hours of actual patient intervention. Additional learning is facilitated by chatbots and virtual assistants that are powered by AI and port their users with answers to queries, guide them through case studies and give instant feedback to the students as to how they are progressing.

Continuing medical education is one length of research that is being conducted by AI, which continues to keep healthcare professionals updated with latest medical research and treatment protocols. Through analyzing huge datasets from medical journals, trials, records, etc., AI empowers physicians with up-to-date, evidence-based insights relevant to a physician's specialization. It makes sure that medical professionals keep up with the advances in medicinal drugs, ailments and innovative treatments.

While AI in medical education brings with it a lot of advantages, there are issues to consider ethical issues, privacy of data, and how to validate AI-generated content. The key is ensuring that the way AI backs up rather than replaces traditional medical training methods maintains the level of education and human-cantered learning.

Continued development of AI will enhance its involvement in medical education and training, which will become more interactive, efficient, and available, as learning evolves with the development of AI. Using AI, medical institutions can ramp up the skills and readiness of future healthcare providers to improve the outcomes for patients and to advance medicine.

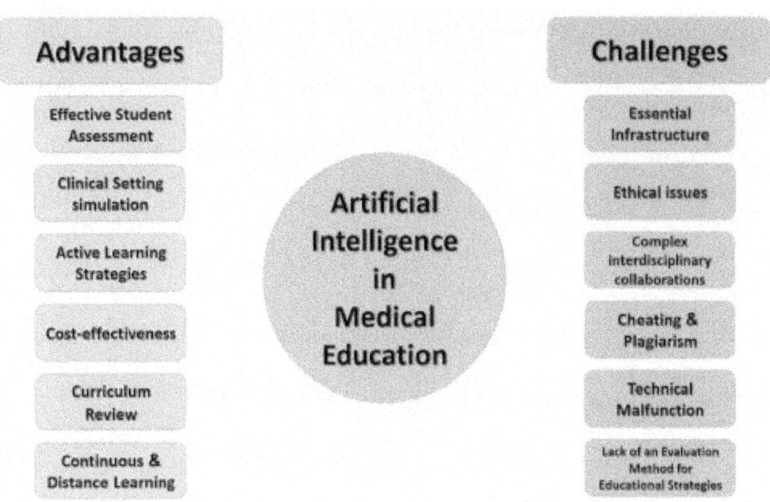

Figure 5.7: The advantages and challenges of the application of artificial intelligence in medical education.

Source: (Zarei et al., 2024)

The graphic symbolizes a mind map that is focused on "Artificial Intelligence in Medical Education," describing the benefits and obstacles associated with it. The advantages, presented in green boxes on the left, include effective student assessment, clinical setting simulation, active learning strategies, cost-effectiveness, curriculum review, and continuous & distance learning. Conversely, the challenges, shown in orange boxes on the right, encompass the need for essential infrastructure, ethical issues, complex interdisciplinary collaborations, the risk of cheating & plagiarism, technical malfunction, and the lack of an evaluation method for educational strategies.

AI in Medical Education Advantages

There are a lot of potential upsides to incorporating AI into medical education, such as better curriculum design and evaluation and the chance to apply novel teaching strategies in clinical settings. The incorporation of AI into medical education has become an indisputable fact, and this section will examine some of the benefits that have contributed to this.

1. There has been a rise in the use of VR alongside the implementation of AI in medical education. With its inventive, efficient, and cost-effective solutions, virtual reality (VR) has the potential to revolutionize medical education in areas like anatomy and surgery. It can also hasten and improve student instruction.

2. AI can streamline the process of formative and summative assessment, which in turn allows for more personalized coaching and a better learning environment for students through the provision of feedback.

3. The use of clinical scenario simulators allows students to practice medical reasoning in a safe environment while also allowing them to learn from their mistakes. Improve your students' diagnostic skills by a significant margin with these AI-powered tools.

4. Transparency is enhanced and valid comparisons of medical education effectiveness across nations and institutions are made possible through the use of AI in education. Additionally, it has the ability to inspire both individuals and educational institutions to own up to their mistakes.

5. Distance learning is made possible by integrating AI into medical education, which opens up medical education to communities with limited resources and remote locations. This can greatly assist recently graduated doctors while they are delivering care. This benefit guarantees continuous student-centered learning even in the face of catastrophic catastrophes like the COVID-19 pandemic.

6. Various active learning methodologies, including problem-based, case-based, small-group, and large-group learning, can be aided by artificial intelligence. Determining what the students need and then providing it, improves the standard of medical education. In contrast, AI may one day make it feasible to share rare clinical situations with more students than is currently feasible owing to constraints in both academic and clinical contexts.

5.5 Roadmap for Sustainable AI Adoption in Healthcare

AI adoption in healthcare has the opportunity to rock patient care, speed up clinical workflows as well as improve medical research. Yet, to have an integrated and sustainable AI requires a well-structured approach that includes both technical and ethical, regulatory, and operational challenges. While the road to implementing AI solutions in healthcare institutions is underway, a well-crafted roadmap will lead the course and help them adopt AI solutions, keeping it simple, and safe, and ensuring patient trust (Ahmed et al., 2023).

The starting point of AI adoption is to evaluate the necessities and targets of the healthcare system. Specific challenges that AI can address are to help improve diagnostics, enhance predictive analytics, develop drug discovery and automate administrative tasks. The gap analysis can be utilized to determine the healthcare institutions' Technological Readiness and the availability of resources required to adopt AI, the areas where it will yield the most impact.

The proper formation of an AI governance framework to ensure ethical and regulatory compliance to construct a strong AI. To ensure proper use of AI in healthcare, healthcare organizations have to have clear policies on how to use AI, data security, accountability, and so on while adhering to regulations such as HIPAA, GDPR. That is why the questions about transparency in AI decision making still feel extremely important: maintaining trust in the medical professionals who are making decisions based upon these models is vital. Also, early involvement of clinical, hospital, policy, and patient stakeholders in the adoption of AI can allay fears about the displacement of jobs, trust in AI systems, and ethical issues, and make AI seen as an augment (rather than a replacement).

To be the successful deployment of the AI driven healthcare solutions, a robust AI infrastructure is required. Since AI systems are not mutually connected, the EHR services (for example, electronic health records) and other hospital IT frameworks are going to have to change to become interoperable with high-performance computing, cloud storage used by the AI systems. This integration ensures no disruption to the existing workflow by allowing AI applications to operate efficiently in the same workflow.

Data management, of course, is equally important as it is to ensure proper quality, and secureness. Medical data must be made standardized so that it improves the training of the AI, as well as implement strict data encryptions, access control, and cybersecurity measures to maintain patient privacy. Fairness and accuracy in AI-driven decision-making and addressing biases in AI models are equally important, which need the generation of diverse, de-identified, and ethically sourced datasets to avoid all forms of biases. AI models must go through a thorough validation and pilot test before they can be deployed in a full scale, and it must be verified that it can be relied on and be safe in real-world medical scenarios.

In developing the AI Maturity Roadmap, three goals were considered:

- Make it easier for businesses to gauge their current level of AI development and plan for the future

- Take all relevant human, technological, and process aspects into account

- When it comes to matters like vendor strategy, it's crucial to find the sweet spot between being opinionated and being flexible.

The six primary components of the roadmap were established in the first workstream: information architecture, culture, governance, business implementation, value, and maintenance and operations. To ensure the long-term success of AI deployment, the AI Collaborative has identified the following essential components.

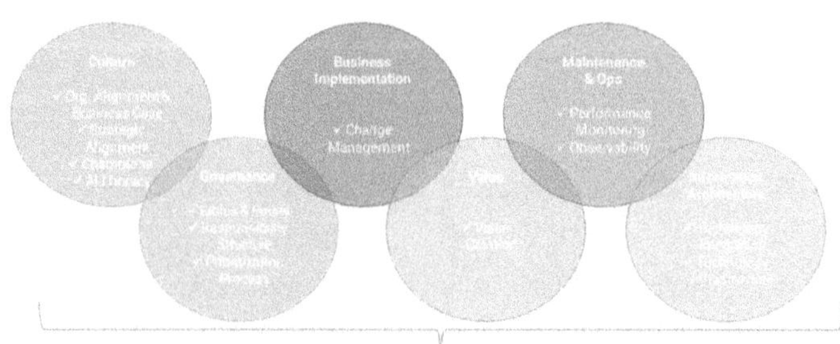

Figure 5.8: Defining Maturity in the AI Space: Evaluation Areas.

Source: (Durlach et al., 2024)

Several important areas that health systems should prioritize are organized by the top-level pillars, which act as general themes. For instance, health systems should prioritize AI literacy, champions, organizational business cases, strategic alignment, and organizational alignment within Culture. Ethical and equitable considerations, accountability structures, and the prioritization process receive further attention in governance.

Making good use of AI requires more than just a change in technology; in fact, most healthcare organizations will need to undergo a significant culture transition in tandem with any AI projects. AI is quickly becoming more than just a data scientist or evangelist tool; as a result, more people in the company should learn how to implement, oversee, and get the most out of AI in their own roles. With a workforce that is more knowledgeable about AI, there will be more people who can advocate for AI efforts in the health system, more people who can own projects, and better alignment among decision-makers.

Health systems that priorities technology above people should reevaluate their governance structure to ensure it is suitable for the future of AI, according to the AI Collaborative, which argues that good governance is a prerequisite for AI maturity.

Prior to integrating AI, health institutions should set up well-defined standards, evaluation procedures, and accountable parties; nonetheless, the entire sector still needs to address established principles of equality and ethics. Crucial concerns regarding who is involved in conducting equity and ethics evaluations, when is human intervention necessary, how can data and people be adequately protected, and how can transparency be maintained should be addressed by this framework. Effective AI initiative teams will work with their health system's current procedures to align expectations, deadlines, and approval criteria for things like technical architecture review, ethics assessment, and risk appraisal.

Three Uses for the AI Maturity Roadmap

The roadmap description permits multiple uses of the matrix system in a health system context. The AI Collaborative designed their framework to be flexible and widely usable because various organizations need to decide their AI maturity path according to their specific needs.

Every health system leader must create their own approach when utilizing the matrix framework according to their organizational needs. Organizations apply it as:

- **A blueprint for maturity:** Health system leaders have the option of following the roadmap in a sequential fashion, with each component and level serving as a guide to the optimal path to reach maturity. This method can help health systems that start at Level 1 generally or those that intend to start at a higher level with their initiatives. They are free to use it in general or to supplement their work on a particular subject (like Governance, for example).

- **A compass for AI:** As a navigational aid, the matrix can show health systems the route without prescribing specific methods; it can only serve as a guide. The document will be updated over time to include validated concepts and best practices for using AI in healthcare contexts.

- **A communication tool:** The success of AI deployments is highly dependent on the level of alignment between C-suite executives, the team responsible for implementing the technology, and the end users. The matrix can serve as a useful tool for these groups to use in their discussions, bolstering the case for investing in the technology and the larger cultural shifts that are necessary.

It is the fundamental belief of the AI Collaborative that most organizations will most likely not see linear improvement. As their requirements and the AI environment change, they could suffer obstacles or completely skip steps. This real implementation of the roadmap, which may vary across health systems, will be crucial to further revisions and enhancements.

5.5.1 Green AI for Energy-Efficient Medical Models

There has been a very rapid integration of artificial intelligence (AI) into health care, and this has enabled a very rapid advancement of diagnostics, treatment planning and operational efficiency in the health care system. While these AI models can be trained with huge computational power needed for such jobs very often consumes high energy and leaves a carbon footprint. In the medical field, the practice of developing AI models with high performance but with reduced energy consumption is the concept of Green AI, which is becoming important. Since sustainability is increasingly a priority, having energy efficient AI models in the healthcare industry means areas of medical innovation continue to inspire reductions in environmental impact without sacrificing accuracy and efficiency (Alowais et al., 2023)but artificial intelligence (AI.

The Need for Energy-Efficient AI in Healthcare

Deep learning algorithms require a lot of large data to be processed and a huge amount of computational resources. Deep neural networks used in such AI mediated medical applications as radiology image analysis, genomic sequencing, predictive analytics are dependent on GPUs of high performing power and cloud computing infrastructure. The increase in energy consumption corroborated by the increased environmental impact of AI driven in the healthcare solution due to this computational demand not only makes access to AI empowered healthcare more expensive, but it also discourages its use, if not outright.

Green AI refers to the optimization of AI models for the outcome of high energy efficiency without compromising performance. Applying energy efficient AI models for medical applications may save operational costs, lessen in the carbon emissions, and make AI-powered healthcare more accessible in resource constraint settings. In the race to achieve sustainability goals, Green AI integration as hospitals and research institutions commences the integration of Green AI in order to not compromise technological advancement in environmental responsibility.

Strategies for Implementing Green AI in Healthcare

- **Model Optimization and Efficiency:** Model compression and pruning is one of the key approaches of Green AI where models are made smaller without losing their abilities of prediction. AI systems that process medical data consume less energy due to techniques of quantization, distilled knowledge, and sparsity aware training.

- **Federated Learning and Edge Computing:** The classic cloud AI model is well based on big centers but uses a lot of energy for storing and registration data. Fed Reached learning enables training of the AI models over multiple local devices without need to move the data to central server to thereby save energy-intensive data transmission. Another common application of edge computing is the use of real-time AI analysis on medical devices, wearable sensors, hospital equipment, and so on, which minimizes the need for application in the cloud and increases energy efficiency.

- **Renewable Energy Integration in AI Infrastructure:** If hospitals and research institutions use the solutions based on AI, the carbon footprint of data centers can be reduced by providing them with renewable energy sources. AI infrastructure from the solar or wind can power medical imaging analysis, disease prediction model, as well as robotic surgery without impacting the environment.

- **Sustainable Data Processing:** There is an immense amount of data generated in the medical region, like hospital "Electronic Health Records" (EHRs), genomic sequencing, medical imaging, etc. To simplify this while still training the AI model on high-quality, relevant datasets without burning unnecessary computational resources, there are efficient data handling techniques, namely secure data lossless compression, selective data retention, and smart data sampling.

Benefits of Green AI in Medical Applications

Lower Operational Costs: Working with more energy-efficient AI models, computing expenses of AI-based medical solutions decrease further, resulting in reduced costs of such solutions for hospitals and clinics.

Sustainable Healthcare Innovation: Green AI has been created to make sure that future medical AI has nothing but good with regards to global sustainability goals and responsible technology adoption.

Increased Accessibility: Models for energy efficient AI driven healthcare also enables the deployment of such healthcare in resource limited environments like rural hospitals in developing countries without requiring the high-power infrastructure.

Ethical AI Deployment: Green AI aligns ethically with ethical AI principles by lowering environmental footprint to support human wellbeing as well as planetary health through health care innovations.

5.5.2 Ensuring Long-Term AI Sustainability in Healthcare

AI integration into healthcare is changing the patient care, diagnostics, and operations in the hospitals. But for AI to be sustainable ethical and environmental in the long run, it must be sustainable. To build long-term AI sustainability in healthcare, that approach needs to combine technological advancement, energy efficiency, regulatory compliance, and financial feasibility. If these aspects of AI systems are absent, they may be both unreliable and costly (or even unethically problematic) and will severely limit their long-term benefits.

If in development and deployment, AI needs to be responsible. Caution has to be taken such that the ethical issues related to bias in AI algorithms, lack of transparency and displacement of jobs are addressed. Medical professionals should not be replaced by AI models, but instead enhanced by them, and decision-making should never be removed from human control. Moreover, these are explained ability and interpretability; clinicians should be able to understand how AI arrives at its recommendations. As a result of the bias in medical data and rules for evaluating medical outcomes, strategies like training AI with varied data and auditing predictions periodically may mitigate the chances of discriminating against patients.

There are tons of patient data required for AI in healthcare, and therefore security and privacy protection is top priorities. The foundation of preserving trust is rigorous adherence to laws like the General Data Protection Regulation (GDPR), the Health Insurance Portability and Accountability Act (HIPAA), and any other local data protection legislation. To minimize risks, healthcare institutions need to implement end-to-end encryption, federated learning techniques as well as access controls. Federated learning allows to train AI models at multiple healthcare centers without having to share raw patient data which substantially helps alleviate privacy concerns but keep the system accurate.

High energy consumption and environmental impact is needed for AI systems due to the require of significant computing power. To achieve AI sustainability, Green AI practices must be used in the creation of air models, such that energy efficiency optimization is applied without downsizing model performance. To tackle the problem of energy use issues due to low computational complexity, techniques like model pruning, quantization, and knowledge distillation are used. It also could reduce the power used by the servers that cloud processing would consume. Edge computing and federated learning could use AI to process the data on medical devices or in the cloud rather than on cloud servers. Further, by integrating renewable energy sources, healthcare institutions can reduce even further their carbon footprint and the AI-powered infrastructure can help them do so.

In order for AI to stay effective at any time, it has to keep learning and learning. Medical knowledge changes constantly, and AI models have to be trained to comply with new treatment recommendations, new conditions or with changing demographics of the patient population. Real time learning mechanisms which allow for the models to improve with new patient outcomes and medical research are important in having sustainable AI systems. This facility lends itself to the potential for AI to stay ahead of the curve and use AI in helping healthcare professionals make the correct diagnosis and treatment plan. Also, importantly, collaboration between the medical practitioners and the AI developer is necessary for refinements needed for clinical uses of AI tools.

The implementation of AI in healthcare must be financially viable too. However, for example, small healthcare institutions can make

AI infrastructure and training high costs of barriers, especially if they choose to finance it on their own. This kind of hospital AI services model, AI as a Service (Alas), allows hospitals to utilize the power of AI by not spending from the get go. It also helps with the automation of administrative tasks by people, monitoring of patient's health, and medical documentation to reduce the cost of operation and make healthcare system more efficient on the base of AI. That did not mean that the importance of training medical professionals to effectively use AI didn't exist lack of training may cause AI tools to be underutilized or misinterpreted, thereby minimizing all their effects.

Multiple Choice Questions (MCQs)

1. **What is a key benefit of explainable and interpretable AI in healthcare?**

   a. Increased computational speed

   b. Improved patient trust and understanding of AI decisions

   c. Reduced need for healthcare professionals

   d. Enhanced AI autonomy without human oversight

2. **How is quantum computing expected to impact drug discovery?**

   a. By increasing the cost of drug production

   b. By slowing down the analysis of complex molecules

   c. By accelerating molecular simulations and identifying new compounds

   d. By replacing traditional chemical research methods entirely

3. **Which of the following is an example of AI in preventive healthcare?**

   a. Robotic-assisted surgery

   b. AI-driven predictive health analytics

   c. Automated medical billing systems

   d. Virtual reality training for surgeons

4. **How can AI improve early diagnosis in cancer screening?**

    a. By increasing radiation doses

    b. By analyzing patterns in medical imaging for early signs of cancer

    c. By recommending alternative treatments automatically

    d. By replacing oncologists in the diagnostic process

5. **What is a major challenge in using AI to bridge healthcare gaps in rural areas?**

    a. Lack of patient willingness to use AI

    b. High availability of medical infrastructure

    c. Limited access to internet and technological resources

    d. Over-reliance on human medical professionals

6. **AI-powered affordable healthcare solutions primarily aim to:**

    a. Increase healthcare costs for advanced treatments

    b. Improve access to basic and advanced medical care

    c. Make healthcare more reliant on human intervention

    d. Replace existing healthcare infrastructure

7. **In what way can AI assist medical professionals in clinical practice?**

    a. By automating all medical procedures

    b. By providing real-time diagnostic insights and decision support

    c. By reducing the need for human involvement in healthcare

    d. By discouraging the use of human judgment in treatment

8. **How is AI enhancing medical education and training?**

    a. By automating patient care

    b. By providing personalized simulations and performance feedback

    c. By increasing reliance on traditional medical books

    d. By eliminating the need for human instructors

9. **What is the goal of Green AI in healthcare?**

   a. To increase computational complexity in medical research

   b. To maximize energy consumption for faster model training

   c. To develop energy-efficient AI models for medical applications

   d. To create AI models without considering environmental impact

10. **Ensuring long-term AI sustainability in healthcare requires:**

   a. Increasing the computational cost of AI models

   b. Developing AI models without considering scalability

   c. Creating infrastructure that supports low-energy AI processing

   d. Reducing AI usage in healthcare entirely

Answers:

1	2	3	4	5	6	7	8	9	10
b	c	b	b	c	b	b	b	c	c

Bibliography

Abayomi, S., Modupe, T., Obioha, O., Omolara, O., & Samuel, O. (2024). *AI - Powered Information Governance : Balancing Automation and Human Oversight for Optimal Organization Productivity. 17*(10), 110–131.

Ahmed, M. I., Spooner, B., Isherwood, J., Lane, M., Orrock, E., & Dennison, A. (2023). A Systematic Review of the Barriers to the Implementation of Artificial Intelligence in Healthcare. *Cureus*, *15*(10), e46454. https://doi.org/10.7759/cureus.46454

Ahsan, M., Khan, A., Khan, K., Sinha, B., & Sharma, A. (2023). Advancements in medical diagnosis and treatment through machine learning: A review. *Expert Systems*, *41*. https://doi.org/10.1111/exsy.13499

Albahri, A. S., Duhaim, A. M., Fadhel, M. A., Alnoor, A., Baqer, N. S., Alzubaidi, L., Albahri, O. S., Alamoodi, A. H., Bai, J., Salhi, A., Santamaría, J., Ouyang, C., Gupta, A., Gu, Y., & Deveci, M. (2023). A systematic review of trustworthy and explainable artificial intelligence in healthcare: Assessment of quality, bias risk, and data fusion. *Information Fusion*, *96*, 156–191. https://doi.org/10.1016/j.inffus.2023.03.008

Alexiei Dingli, D. F. (2023). *Neuro-Symbolic AI.*

Alivisatos, A. P., Andrews, A. M., Boyden, E. S., Chun, M., Church, G. M., Deisseroth, K., Donoghue, J. P., Fraser, S. E., Lippincott-Schwartz, J., Looger, L. L., Masmanidis, S., McEuen, P. L., Nurmikko, A. V., Park, H., Peterka, D. S., Reid, C., Roukes, M. L., Scherer, A., Schnitzer, M., … Zhuang, X. (2013). Nanotools for neuroscience and brain activity mapping. *ACS Nano*, *7*(3), 1850–1866. https://doi.org/10.1021/nn4012847

Alowais, S. A., Alghamdi, S. S., Alsuhebany, N., Alqahtani, T., Alshaya, A. I., Almohareb, S. N., Aldairem, A., Alrashed, M., Bin Saleh, K., Badreldin, H.

A., Al Yami, M. S., Al Harbi, S., & Albekairy, A. M. (2023). Revolutionizing healthcare: the role of artificial intelligence in clinical practice. *BMC Medical Education*, *23*(1), 689. https://doi.org/10.1186/s12909-023-04698-z

Amiri, Z., Heidari, A., Darbandi, M., Yazdani, Y., Jafari Navimipour, N., Esmaeilpour, M., Sheykhi, F., & Unal, M. (2023). The Personal Health Applications of Machine Learning Techniques in the Internet of Behaviors. *Sustainability*, *15*(16), 12406. https://doi.org/10.3390/su151612406

Askin, S., Burkhalter, D., Calado, G., & El Dakrouni, S. (2023). Artificial Intelligence Applied to clinical trials: opportunities and challenges. *Health and Technology*, *13*(2), 203–213. https://doi.org/10.1007/s12553-023-00738-2

Bajwa, Munir, J., Nori, U., & Williams, A. (2021). Artificial intelligence in healthcare: transforming the practice of medicine. *Future Healthcare Journal*, *8*(2), e188–e194. https://doi.org/10.7861/fhj.2021-0095

Batool, A., & Lopez, A. (2023). *Healthcare Access and Regional Connectivity: Bridging the Gap*. *02*, 260–271.

Bellec, P., & Boyle, J. (2019). Bridging the gap between perception and action: the case for neuroimaging, AI and video games. *PsyArXiv*.

Bleher, H., & Braun, M. (2022). Diffused responsibility: attributions of responsibility in the use of AI-driven clinical decision support systems. *AI and Ethics*, *2*. https://doi.org/10.1007/s43681-022-00135-x

Blobel, B. (2002). *Analysis, Design and Implementation of Secure and Interoperable Distributed Health Information Systems*.

Boppana, V. R. (2019). *Role of IoT in Remote Patient Monitoring Systems*. *2*, 1–25.

Carrasco Ramírez, J. G. (2024). AI in Healthcare: Revolutionizing Patient Care with Predictive Analytics and Decision Support Systems. *Journal of Artificial Intelligence General Science (JAIGS) ISSN:3006-4023*. https://doi.org/10.60087/jaigs.v1i1.p37

Chakraborty, C., Bhattacharya, M., Pal, S., & Lee, S. S. (2024). From machine learning to deep learning: Advances of the recent data-driven paradigm shift in medicine and healthcare. In *Current Research in Biotechnology*. https://doi.org/10.1016/j.crbiot.2023.100164

Colonna, L. (2021). Artificial Intelligence in the Internet of Health Things: Is the Solution to AI Privacy More AI? *SSRN Electronic Journal.* https://doi.org/10.2139/ssrn.3838571

D. Satishkumar, M. S. (2024). *Machine learning has revolutionized many industries by automating decision-making processes and improving efficiency.*

Daly, A., Hagendorff, T., Li, H., Mann, M., Marda, V., Wagner, B., & Wang, W. W. (2020). AI, Governance and Ethics: Global Perspectives. *SSRN Electronic Journal.* https://doi.org/10.2139/ssrn.3684406

Desarno, J., Perez, M., Rivas, R., Sandate, I., Reed, C., & Fonseca, I. (2021). Succession Planning Within the Health Care Organization:: Human Resources Management and Human Capital Management Considerations. *Nurse Leader.* https://doi.org/10.1016/j.mnl.2020.08.010

Dinh-Le, C., Chuang, R., Chokshi, S., & Mann, D. (2019). Wearable Health Technology and Electronic Health Record Integration: Scoping Review and Future Directions. *JMIR MHealth and UHealth, 7*(9), e12861. https://doi.org/10.2196/12861

Dixon, D., Sattar, H., Moros, N., Kesireddy, S. R., Ahsan, H., Lakkimsetti, M., Fatima, M., Doshi, D., Sadhu, K., & Junaid Hassan, M. (2024). Unveiling the Influence of AI Predictive Analytics on Patient Outcomes: A Comprehensive Narrative Review. *Cureus, 16*(5), e59954. https://doi.org/10.7759/cureus.59954

Durlach, P., Fournier, R., Gottlich, J., Markwell, T., McManus, J., Merrill, A., & Rhew, D. (2024). The AI Maturity Roadmap: A Framework forEffective and Sustainable AI in Health Care. In *Nejm Ai* (Vol. 1, Issue 4, pp. 1–6).

Edemekong, P. F., Annamaraju, P., Afzal, M., & Haydel, M. J. (2025). *Health Insurance Portability and Accountability Act (HIPAA) Compliance.* StatPearls Publishing, Treasure Island (FL).

Elendu, C., Amaechi, D. C., Okatta, A. U., Amaechi, E. C., Elendu, T. C., Ezeh, C. P., & Elendu, I. D. (2024). The impact of simulation-based training in medical education: A review. *Medicine, 103*(27), e38813. https://doi.org/10.1097/MD.0000000000038813

Evans, R. S. (2016). Electronic Health Records: Then, Now, and in the Future. *Yearbook of Medical Informatics*, S48–S61. https://doi.org/10.15265/IYS-2016-s006

Finnerty, K. T. (2024). *Representation Matters: Elevating Diverse Voices in a Predominantly White High School.*

Fisher, J. A. (2009). Medical research for hire: The political economy of pharmaceutical clinical trials. In *Medical Research for Hire: The Political Economy of Pharmaceutical Clinical Trials.* https://doi.org/10.1177/009430610903800530

George, A. S. H., Shahul, A., & George, D. A. S. (2023). Artificial {Intelligence} in {Medicine}: {A} {New} {Way} to {Diagnose} and {Treat} {Disease}. *Partners Universal International Research Journal, 2*(3), 246–259. https://doi.org/10.5281/zenodo.8374066

Gerdin, J. (2021). *Health Careers Today E-Book.*

Gumbs, A., De Simone, B., & Chouillard, E. (2020). Searching for a better definition of robotic surgery: is it really different from laparoscopy? *Mini-Invasive Surgery, 4.* https://doi.org/10.20517/2574-1225.2020.110

Gupta, R., Srivastava, D., Sahu, M., Tiwari, S., Ambasta, R. K., & Kumar, P. (2021). Artificial intelligence to deep learning: machine intelligence approach for drug discovery. *Molecular Diversity.* https://doi.org/10.1007/s11030-021-10217-3

Gutierrez, A. R. (2023). EXPLORING THE FUTURE OF PROSTHETICS AND ORTHOTICS: HARNESSING THE POTENTIAL OF 3D PRINTING. *Canadian Prosthetics and Orthotics Journal.* https://doi.org/10.33137/cpoj.v6i2.42140

Halabi, M., Khoury, K., Alomar, A., Dahdah, J. El, Hassan, O., Hayyan, K., Bishara, E., & Moussa, H. (2024). Operative efficiency: a comparative analysis of Versius and da Vinci robotic systems in abdominal surgery. *Journal of Robotic Surgery.* https://doi.org/10.1007/s11701-023-01806-5

Hno, T. E. C., & Gy, L. O. (2016). Virtual Reality ' S Lab Appeal. *Nature, 533,* 153.

Hunter, B., Hindocha, S., & Lee, R. W. (2022). The Role of Artificial Intelligence in Early Cancer Diagnosis. *Cancers, 14*(6), 1524. https://doi.org/10.3390/cancers14061524

Iqbal, F. (2025). *ADMINISTRATIVE COSTS IN THE US HEALTHCARE SYSTEM : IS SIMPLIFICATION AND TECHNOLOGY A SOLUTION TO EFFICIENCY AND GROWING HEALTHCARE ADMINISTRATIVE COSTS IN THE US HEALTHCARE SYSTEM : IS SIMPLIFICATION AND TECHNOLOGY A SOLUTION TO EFFICIENCY AND GROWING HEALTHCARE COSTS ? February.* https://doi.org/10.5281/zenodo.14898792

Javaid, M., Haleem, A., Pratap Singh, R., Suman, R., & Rab, S. (2022). Significance of machine learning in healthcare: Features, pillars and applications. *International Journal of Intelligent Networks, 3,* 58–73. https://doi.org/10.1016/j.ijin.2022.05.002

Jin, J. (2018). *Electromagnetic Analysis and Design in Magnetic Resonance Imaging.* Routledge. https://doi.org/10.1201/9780203758731

Kaplan, R. S., & Anderson, S. R. (2003). Time-Driven Activity-Based Costing Robert S. Kaplan and Steven R. Anderson November 2003. *Harvard Business Review, 82*(November), 131–138.

Kaufmann, W. (2017). *Going by the Book.* Routledge. https://doi.org/10.4324/9780203790557

Khang, A. (2024). *Driving Smart Medical Diagnosis Through AI-powered Technologies and Applications.*

Khezr, S., Moniruzzaman, M., Yassine, A., & Benlamri, R. (2019). Blockchain Technology in Healthcare: A Comprehensive Review and Directions for Future Research. *Applied Sciences, 9*(9), 1736. https://doi.org/10.3390/app9091736

Kibira, D., Lee, Y. T., Marshall, J., Feeney, A. B., Avery, L., & Jacobs, A. (2015). Simulation-based design concept evaluation for ambulance patient compartments. *Simulation, 91*(8), 691–714. https://doi.org/10.1177/0037549715592716

Kirilenko, I. (2020). What Is Artificial Intelligence in Software Testing? In *Parasoft*.

Krishna, A. N., Anitha, A. C., & Naveena, C. (2022). Chatbot-An Intelligent Virtual Medical Assistant. *Communications in Computer and Information Science*. https://doi.org/10.1007/978-3-031-22405-8_9

Kromrey, M. L., Steiner, L., Schön, F., Gamain, J., Roller, C., & Malsch, C. (2024). Navigating the Spectrum: Assessing the Concordance of ML-Based AI Findings with Radiology in Chest X-Rays in Clinical Settings. *Healthcare (Switzerland)*, *12*(22). https://doi.org/10.3390/healthcare12222225

Lee, I. R. (2021). Deep Medicine. *Korean Medical Education Review*. https://doi.org/10.17496/kmer.2021.23.1.68

Lhoest, L., Lamrini, M., Vandendriessche, J., Wouters, N., da Silva, B., Chkouri, M. Y., & Touhafi, A. (2021). MosAIc: A Classical Machine Learning Multi-Classifier Based Approach against Deep Learning Classifiers for Embedded Sound Classification. *Applied Sciences*, *11*(18), 8394. https://doi.org/10.3390/app11188394

Li, Y. H., Li, Y. L., Wei, M. Y., & Li, G. Y. (2024). Innovation and challenges of artificial intelligence technology in personalized healthcare. *Scientific Reports*, *14*(1), 1–9. https://doi.org/10.1038/s41598-024-70073-7

Louise, J. (2011). Practical Decision Making in Health Care Ethics: Cases and Concepts (3ed). *Australian and New Zealand Journal of Public Health*, *35*(4), 398. https://doi.org/10.1111/j.1753-6405.2011.00752.x

Maibaum, A., Bischof, A., Hergesell, J., & Lipp, B. (2022). A critique of robotics in health care. *AI and Society*. https://doi.org/10.1007/s00146-021-01206-z

Mammen, J. R., Elson, M. J., Java, J. J., Beck, C. A., Beran, D. B., Biglan, K. M., Boyd, C. M., Schmidt, P. N., Simone, R., Willis, A. W., & Dorsey, E. R. (2018). Patient and Physician Perceptions of Virtual Visits for Parkinson's Disease: A Qualitative Study. *Telemedicine and E-Health*, *24*(4), 255–267. https://doi.org/10.1089/tmj.2017.0119

Matthew Lungren, MD, M. (2024). *Unlocking next-generation AI capabilities with healthcare AI models.*

Mehrabi, N., Morstatter, F., Saxena, N., & Jan, L. G. (2024). *A Survey on Bias and Fairness in Machine Learning.*

Mohamadipanah, H., Perumalla, C., Yang, S., Wise, B., Kearse, L. D., Goll, C., Witt, A., Korndorffer, J. R., & Pugh, C. (2022). Artificial intelligence in surgery: A research team perspective. *Current Problems in Surgery*, *59*(6). https://doi.org/10.1016/j.cpsurg.2022.101125

Mosaiyebzadeh, F., Pouriyeh, S., Parizi, R. M., Sheng, Q. Z., Han, M., Zhao, L., Sannino, G., Ranieri, C. M., Ueyama, J., & Batista, D. M. (2023). Privacy-Enhancing Technologies in Federated Learning for the Internet of Healthcare Things: A Survey. *Electronics (Switzerland).* https://doi.org/10.3390/electronics12122703

Nawaf Mansour Saeed AlQahtani, Saad Mohammed Abdulaziz Alsaaran., E. A. E. A.-M. (2023). *INNOVATING LIFE: THE FUTURE OF BIOMEDICAL ENGINEERING.*

Niu, T., Lei, B., Guo, L., Fang, S., Li, Q., Gao, B., Yang, L., & Gao, K. (2023). A Review of Optimization Studies for System Appointment Scheduling. *Axioms*, *13*(1), 16. https://doi.org/10.3390/axioms13010016

Panesar, A. (2019). Machine Learning and AI for Healthcare. In *Machine Learning and AI for Healthcare.* https://doi.org/10.1007/978-1-4842-3799-1

Patrinos, G. P., & Mitropoulou, C. (2022). Horizon Scanning: Teaching Genomics and Personalized Medicine in the Digital Age. *OMICS A Journal of Integrative Biology.* https://doi.org/10.1089/omi.2021.0119

Peek, N., Combi, C., Marin, R., & Bellazzi, R. (2015). Thirty years of artificial intelligence in medicine (AIME) conferences: A review of research themes. *Artificial Intelligence in Medicine*, *65*(1), 61–73. https://doi.org/10.1016/j.artmed.2015.07.003

Pepper, J. (2023). *The Electronic Health Record for the Physician's Office E-Book For SimChart for the Medical Office.*

Pezoulas, V., Exarchos, T., & Fotiadis, D. (2020). Medical data sharing, harmonization and analytics. In *Medical Data Sharing, Harmonization and Analytics.* https://doi.org/10.1016/c2018-0-00523-x

Rief, S. F. (2005). How to reach and teach children with ADD/ADHD : practical techniques, strategies, and interventions. In *Jossey-Bass teacher*.

Saikia, A., Mazumdar, S., Sahai, N., Paul, S., Bhatia, D., Verma, S., & Rohilla, P. K. (2016). Recent advancements in prosthetic hand technology. *Journal of Medical Engineering and Technology, 40*(5), 255–264. https://doi.org/10.3109/03091902.2016.1167971

Sajja, P. S., & Akerkar, R. (2016). Intelligent Technologies for Web Applications. In *Intelligent Technologies for Web Applications*. https://doi.org/10.1201/b12118

Saqib, M., Iftikhar, M., Neha, F., Karishma, F., & Mumtaz, H. (2023). Artificial intelligence in critical illness and its impact on patient care: a comprehensive review. *Frontiers in Medicine, 10*(April), 1–8. https://doi.org/10.3389/fmed.2023.1176192

Schelter, S., & Stoyanovich, J. (2022). Taming Technical Bias in Machine Learning Pipelines *. *Ssc.Io, 1926250*, 1–12.

Scholz, M. L., Collatz-Christensen, H., Blomberg, S. N. F., Boebel, S., Verhoeven, J., & Krafft, T. (2022). Artificial intelligence in Emergency Medical Services dispatching: assessing the potential impact of an automatic speech recognition software on stroke detection taking the Capital Region of Denmark as case in point. *Scandinavian Journal of Trauma, Resuscitation and Emergency Medicine, 30*(1), 1–17. https://doi.org/10.1186/s13049-022-01020-6

Secinaro, S., Calandra, D., Secinaro, A., Muthurangu, V., & Biancone, P. (2021). The role of artificial intelligence in healthcare: a structured literature review. *BMC Medical Informatics and Decision Making, 21*(1), 125. https://doi.org/10.1186/s12911-021-01488-9

Sendak, M., Elish, M. C., Gao, M., Futoma, J., Ratliff, W., Nichols, M., Bedoya, A., Balu, S., & O'Brien, C. (2020). "The Human Body is a Black Box": Supporting Clinical Decision-Making with Deep Learning. *FAT* 2020 - Proceedings of the 2020 Conference on Fairness, Accountability, and Transparency*, 99–109. https://doi.org/10.1145/3351095.3372827

Sendelj, R., & Ognjanovic, I. (2022). Cybersecurity Challenges in Healthcare. *Studies in Health Technology and Informatics, 300*, 190–202. https://doi.org/10.3233/SHTI220951

Sennaar, K. (2019). *AI in Medical Devices – Three Emerging Industry Applications*. EmeRD.

Shankar, K., Perumal, E., & Gupta, D. (2021). *Artificial Intelligence for the Internet of Health Things*. CRC Press. https://doi.org/10.1201/9781003159094

Shashi Manish. (2022). Digital Strategies to Improve the Performance of Pharmaceutical Supply Chains. *Walden University*.

Shaw, P. L. (2023). *Calculating and Reporting Healthcare Perpustakaan Universitas Esa Unggul Quality and Performance Improvement in Healthcare Page 2 of 6 Artificial Intelligence And Data Mining in Healthcare Two Thousand Twenty Three CDI Pocket Guide Practical Time Series A. 11510*(9), 1–6.

Sonkamble, R. G., Bongale, A. M., Phansalkar, S., Sharma, A., & Rajput, S. (2023). Secure Data Transmission of Electronic Health Records Using Blockchain Technology. *Electronics (Switzerland), 12*(4). https://doi.org/10.3390/electronics12041015

Stoumpos, A. I., Kitsios, F., & Talias, M. A. (2023). Digital Transformation in Healthcare: Technology Acceptance and Its Applications. *International Journal of Environmental Research and Public Health*. https://doi.org/10.3390/ijerph20043407

Subramanian, M., Wojtusciszyn, A., Favre, L., Boughorbel, S., Shan, J., Letaief, K. B., Pitteloud, N., & Chouchane, L. (2020). Precision medicine in the era of artificial intelligence: implications in chronic disease management. *Journal of Translational Medicine, 18*(1), 1–12. https://doi.org/10.1186/s12967-020-02658-5

Tailor, K. (2015). *The Patient Revolution*.

Topol, E. (2019). Deep Medicine - How Artificial Intelligence Can Make Healthcare Human Again. In *Journal of Chemical Information and Modeling*.

Tsvetanov, F. (2024). Integrating AI Technologies into Remote Monitoring Patient Systems. *International Conference on Electronics, Engineering Physics and Earth Science (EEPES 2024)*, 54. https://doi.org/10.3390/engproc2024070054

Udai Pratap Rao, Sweta Gupta, Piyush Kumar Shukla, Chandan Trivedi, Z. S. S. (2021). *Blockchain for Information Security and Privacy*.

Ul, N., Tahir, A., Rashid, U., Hadi, H. J., Ahmad, N., Cao, Y., Alshara, M. A., & Javed, Y. (2024). *Blockchain-Based Healthcare Records Management Framework : Enhancing Security , Privacy , and Interoperability*. 1–19.

Väänänen, A., Haataja, K., Vehviläinen-Julkunen, K., & Toivanen, P. (2021). AI in healthcare: A narrative review. *F1000Research*, *10*, 6. https://doi.org/10.12688/f1000research.26997.2

Van Calster, B., Wynants, L., Timmerman, D., Steyerberg, E. W., & Collins, G. S. (2019). Predictive analytics in health care: how can we know it works? *Journal of the American Medical Informatics Association*, *26*(12), 1651–1654. https://doi.org/10.1093/jamia/ocz130

Vannaprathip, N., Haddawy, P., Schultheis, H., & Suebnukarn, S. (2022). Intelligent Tutoring for Surgical Decision Making: a Planning-Based Approach. *International Journal of Artificial Intelligence in Education*. https://doi.org/10.1007/s40593-021-00261-3

Visibelli, A., Roncaglia, B., Spiga, O., & Santucci, A. (2023). The Impact of Artificial Intelligence in the Odyssey of Rare Diseases. In *Biomedicines*. https://doi.org/10.3390/biomedicines11030887

Wager, K. A., Lee, F. W., & Glaser, J. P. (2009). Health Care Information Systems A practical approach for health care management. In *Hellenic journal of nuclear medicine*.

Westerlund, A. M., Hawe, J. S., Heinig, M., & Schunkert, H. (2021). Risk Prediction of Cardiovascular Events by Exploration of Molecular Data with Explainable Artificial Intelligence. *International Journal of Molecular Sciences*, *22*(19), 10291. https://doi.org/10.3390/ijms221910291

Woods, S. (2006). *EBOOK: Death's Dominion: Ethics at the End of Life*.

Yoon, F. (2021). Artificial Intelligence for Drug Development, Precision Medicine, and Healthcare. *International Statistical Review.* https://doi.org/10.1111/insr.12440

Zafar, F., Fakhare Alam, L., Vivas, R. R., Wang, J., Whei, S. J., Mehmood, S., Sadeghzadegan, A., Lakkimsetti, M., & Nazir, Z. (2024). The Role of Artificial Intelligence in Identifying Depression and Anxiety: A Comprehensive Literature Review. *Cureus.* https://doi.org/10.7759/cureus.56472

Zarei, M., Eftekhari Mamaghani, H., Abbasi, A., & Hosseini, M. S. (2024). Application of artificial intelligence in medical education: A review of benefits, challenges, and solutions. *Medicina Clinica Practica, 7*(2), 7–11. https://doi.org/10.1016/j.mcpsp.2023.100422

Zeleznik, R., Foldyna, B., Eslami, P., Weiss, J., Alexander, I., Taron, J., Parmar, C., Alvi, R. M., Banerji, D., Uno, M., Kikuchi, Y., Karady, J., Zhang, L., Scholtz, J. E., Mayrhofer, T., Lyass, A., Mahoney, T. F., Massaro, J. M., Vasan, R. S., … Aerts, H. J. W. L. (2021). Deep convolutional neural networks to predict cardiovascular risk from computed tomography. *Nature Communications.* https://doi.org/10.1038/s41467-021-20966-2

About the Authors

My name is **Harshit Kohli**, and I am based out of Raleigh, North Carolina, USA. I am currently working with a leading tech company, bringing over 14 years of rich experience in IT infrastructure management and solution design. My career has been marked by a strong track record in enhancing cloud strategies specifically tailored for healthcare clients,  implementing robust analytics solutions, and effectively managing workloads related to Artificial Intelligence. I have received recognition for my ability to train and mentor teams, drive business process improvements, and conduct thorough architecture reviews, all of which play a crucial role in ensuring client success in a competitive landscape. Currently, I am pursuing a Doctorate in Artificial Intelligence at the University of the Cumberlands in the USA, complementing my academic background that includes a Master of Science in Information Technology and a Bachelor of Technology in Electronics and Instrumentation from India. My extensive experience encompasses collaboration with numerous clients on various Analytics and Artificial Intelligence use cases, where I have consistently demonstrated exceptional efforts and achievements. Additionally, I have had the honor of serving as a keynote speaker at several international conferences focused on Analytics, Computing, and Artificial Intelligence. My involvement extends to reviewing research papers for multiple conferences and chairing sessions to ensure their successful organization. Furthermore, I have acted as a judge for various hackathons and prestigious awards, including Globee and Business Intelligence. Overall, my expertise in Data Analytics and Artificial Intelligence not only enhances my professional capabilities but also deepens my understanding of the field.

My name is **Chitrangadaa Singh Kohli** and I am based out of Raleigh, North Carolina USA. I serve as the IT Applications Programmer Lead at a leading Insurance company, where I leverage over 13 years of extensive experience in managing IT infrastructure and designing innovative solutions. Throughout my career, I have established a solid reputation for enhancing business strategies, specifically within the healthcare and insurance industries, successfully implementing robust solutions while adeptly managing Dot Net application workloads. My contributions have been acknowledged through recognition for my proficiency in crafting effective solutions, resolving intricate bugs, and delivering timely fixes for various products. Currently, I am advancing my academic pursuits by working towards a Doctorate in Artificial Intelligence at the University of the Cumberlands in the USA, which builds upon my educational foundation that includes a Master of Science in Data Science and a Bachelor of Technology in Electronics and Instrumentation from India. My broad experience includes collaborating with diverse teams on a multitude of requirements and use cases, where I have consistently showcased exceptional dedication and notable achievements.